Principles of
Nuclear
Chemistry

Essential Textbooks in Chemistry

ISSN: 2059-7738

Orbitals: With Applications in Atomic Spectra
  by Charles Stuart McCaw

Principles of Nuclear Chemistry
  by Peter A C McPherson

Forthcoming

Astrochemistry: From the Big Bang to the Present
  by Claire Vallance

Atmospheric Chemistry: From the Surface to the Stratosphere
  by Grant Ritchie

Problems of Instrumental Analytical Chemistry: A Hands-On Guide
  by J.M. Andrade-Garda, A. Carolsena-Zubieta, M.P. Gómez-Carracedo,
  M.A. Maestro-Saavedra, M.C. Prieto-Blanco, and R.M. Soto-Ferreiro

Essential Textbooks in Chemistry

Principles of
Nuclear
Chemistry

Peter A C McPherson

Belfast Metropolitan College, UK

 **World Scientific**

NEW JERSEY · LONDON · SINGAPORE · BEIJING · SHANGHAI · HONG KONG · TAIPEI · CHENNAI · TOKYO

Published by

World Scientific Publishing Europe Ltd.

57 Shelton Street, Covent Garden, London WC2H 9HE

Head office: 5 Toh Tuck Link, Singapore 596224

USA office: 27 Warren Street, Suite 401-402, Hackensack, NJ 07601

Library of Congress Cataloging-in-Publication Data

Names: McPherson, Peter A. C., author.

Title: Principles of nuclear chemistry / Peter A.C. McPherson
(Belfast Metropolitan College, UK).

Description: New Jersey : World Scientific, 2016. | Series: Essential textbooks in
chemistry | Includes bibliographical references and index.

Identifiers: LCCN 2016033506| ISBN 9781786340504 (hc : alk. paper) |
ISBN 9781786340511 (pbk : alk. paper)

Subjects: LCSH: Nuclear chemistry.

Classification: LCC QD601.3 .M37 2016| DDC 541/.38--dc23

LC record available at https://lccn.loc.gov/2016033506

British Library Cataloguing-in-Publication Data

A catalogue record for this book is available from the British Library.

Desk Editors: Herbert Moses/Mary Simpson

Typeset by Stallion Press
Email: enquiries@stallionpress.com

Printed in Singapore

Preface

There are few words in the English language which can elicit quite as dramatic a response as the word "nuclear." It has become synonymous with weapons of mass destruction, environmental pollution and any number of science fiction scenarios. Of course, the reality is that we are surrounded by nuclear events every day — casting a glance up into the sky we see the result of nuclear reactions within the Earth's Sun which sustains all life on Earth.

Those of us involved in nuclear chemistry have traditionally occupied somewhat of a minority. However, in recent years, the traditional boundaries between the disciplines of science have become increasingly blurred and this is to be commended — we are a community of scientists rather than chemists, physicists, and biologists. This drawing together of scientific disciplines has been at the forefront of the author's mind in writing this book.

The aim of this text is to provide an overview of the theory and application of nuclear chemistry, without necessarily becoming too involved in mathematics. It will be of use to those taking undergraduate courses in chemistry, biochemistry and the applied disciplines; it may also be of use to those working in radiochemistry or involved in radiopharmacy.

I have sought to provide as clear examples as possible when dealing with the mathematical aspects of the subject. End-of-chapter questions which are calculation-based have worked solutions at the end of the text. Also, some additional mathematical information and more lengthy derivations are given in the Appendixes for those who are interested.

The majority of this text has been used in the teaching of nuclear chemistry and medical physics to undergraduates, as well as general chemistry and physical chemistry classes. As is often the case, we learn a great deal from our students' understanding of a subject, and their feedback over the years has been invaluable. I would also like to express my gratitude to my colleagues at Belfast Metropolitan College, Queen's University, and the University of Oxford for answering queries and providing moral support. I have no doubt that some errors will have escaped my notice; eagle-eyed readers are encouraged to contact me with their comments.

As Petronius said in the Satyricon "... nothing pleases everyone: this man gathers thorns, that one roses." Hopefully, you'll find something of use within this text — more roses than thorns.

Peter McPherson

About the Author

Dr Peter McPherson is a Senior Lecturer in Chemistry at Belfast Metropolitan College where he also serves as the Curriculum Manager for Applied Science. Since his early days in industry, he has been involved in the analytical applications of radioisotopes and is currently collaborating in a radiopharmaceuticals project. His teaching interests focus on physical chemistry and he has recently developed an online course in Nuclear Chemistry for undergraduates. He has previously published *Practical Volumetric Analysis* (RSC Publishing).

Contents

Chapter 1

Concepts in Physics

"Those who aspire not to guess and divine, but to discover and know, who propose not to devise, mimic and fabulous worlds of their own, but to examine and dissect the nature of this very world itself, must go to facts themselves for everything."

F. Bacon (1620)

It is impossible to undertake any serious study of nuclear chemistry without a working understanding of key aspects of physics and mathematics. On completion of this chapter and the associated questions, you should:

- Be familiar with Newton's laws of motion and their relationship to work and energy.
- Be able to mathematically describe uniform circular motion and how it relates to the harmonic motion and properties of waves.
- Have an awareness of basic atomic structure and that unstable atoms gain stability through release of different types of radiation.

1.1 Introduction to Classical Physics

Key Point: Classical physics is Newtonian physics which is built on macroscopic observations.

Much of what we consider to be "classical physics" stems from the work of Sir Isaac Newton (1642–1727), James Clerk Maxwell (1831–1879), and Ludwig Boltzmann (1844–1906) (Figure 1.1). Their theories on kinematics, electromagnetism, and thermodynamics, respectively, denominated the scientific landscape for much of the 19th and early 20th centuries.

Figure 1.1 The classical physicists (left to right): Sir Isaac Newton (1642–1727), James Clerk Maxwell (1831–1879), and Ludwig Boltzmann (1844–1906).

Understandably, these theories focused on the measurement of observable phenomena, such as an apple falling from a tree. To verify their theories, these scientists used a branch of mathematics known as calculus to quantify their observations. We will be making extensive use of calculus in our study of nuclear chemistry, especially when discussing the rate of radioactive decay. For this reason, we will revise many of the basic principles of calculus in Chapter 1.

1.1.1 Newton's first law

Key Point: The change in an object's position over time is given by the velocity of the object; the acceleration is the change in velocity over time.

To begin, consider Isaac Newton sitting beneath an apple tree in his garden at Woolsthorpe Manor in Lincolnshire. Newton thought about the rather obvious fact that an apple will remain attached to the tree unless something knocks/pulls it off. He realized that the same idea also applied to an object moving at constant speed — it would continue to move at that fixed speed unless some other force acted upon it. This forms the basis of Newton's first law which was published in his *Philosophiæ Naturalis Principia Mathematica* in 1687. It was not without controversy and Newton's ideas were famously rebuffed by Robert Hooke (1635–1703), a contemporary of Newton, whose competing work on gravitation was ultimately replaced by Newton's theories.

To understand Newton's first law, we must first introduce two terms which are easily confused: *distance* and *displacement*. To illustrate the difference, imagine Newton walking around the rectangular Great Court at Trinity College, Cambridge. If he walks 4 m east, then 2 m south, then 4 m west and 2 m north, he will have travelled a distance of 12 m. However, because he has ended up at his starting position, we say that his displacement is zero. This difference arises from the type of information conveyed by a measurement of distance or displacement. We say that distance is a **scalar quantity** — it only describes the magnitude of the distance travelled. Displacement is a **vector quantity** — it describes the difference between the initial and final points (we say it has magnitude and direction).

Now consider the famous incident of the apple falling from the tree. Suppose Newton timed how long it took an apple to fall from a branch 3.1 m above ground level. If it took 0.8 s, we can calculate the **average speed** of the apple:

$$\text{Average speed} = \frac{\text{Change in distance}}{\text{Change in time}} = \frac{\Delta d}{\Delta t} = \frac{3.1 \text{ m}}{0.8 \text{ s}} = 3.9 \text{ m/s}$$

In this equation, the numerator, Δd, is read as "delta d" and this means the change in distance; similarly, the term in the denominator, Δt, is the change in time. Since we know both the magnitude and direction of the falling apple, we can replace the change in distance with change in displacement, Δs, and calculate the **average velocity**, $\bar{v}$, of the apple:

$$\bar{v} = \frac{\Delta s}{\Delta t} = \frac{-3.1 \text{ m}}{0.8 \text{ s}} = -3.9 \text{ m/s}$$

Notice the minus sign in front of the displacement and the velocity. This tells us that the displacement and resultant velocity are in the downward direction (velocity is therefore a vector quantity).

Now, suppose that Newton was able to measure the displacement of the apple every fraction of a second as it fell. Plotting a graph of displacement (y-axis) against time (x-axis) would produce something like that shown in Figure 1.2. Notice how the shape of the graph changes with time — it is not a straight line, but follows a curve known as a **parabola**. The significance of this is that the velocity of the apple is changing as it falls: at first, it falls relatively slowly (the curve is flatter), but as time progresses, it falls faster (the curve is steeper). When we use $\Delta d/\Delta t$ or $\Delta s/\Delta t$ to calculate the average velocity, this is not taken into account — it assumes a straight line (linear) relationship. If we wanted to know the velocity of the apple during the initial slower stage, we would have to use **instantaneous velocity**, v.

Suppose we wanted to calculate the instantaneous velocity at 0.5 s. First, we could draw a tangent, AB, to the point at $t = 0.5$ s and the gradient of

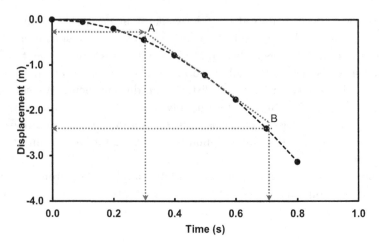

Figure 1.2 Displacement–time graph for an apple in free fall. The gradient of the tangent, AB, can be used to determine the average velocity.

this tangent will equal the instantaneous velocity. For the tangent given in Figure 1.2, the gradient is given by:

$$m_{AB} = \frac{y_2 - y_1}{x_2 - x_1} = \frac{-2.40 - (-0.40)}{0.70 - 0.30} = -5.0\,\text{m/s}$$

The accuracy of the instantaneous velocity determined by this method will depend on the precision with which the tangent to the curve is drawn. The more accurate method involves ***differentiation***, which is a branch of calculus. When using differentiation, we replace the delta symbol Δ, with its Roman equivalent, d. Therefore, our expression for instantaneous velocity becomes

$$v = \frac{ds}{dt} \tag{1.1}$$

Expressions such as Eq. (1.1) are referred to as differentials, and they provide a means of expressing rates of change.

To use Eq. (1.1) to calculate the instantaneous velocity of our falling apple, we must be able to describe the curve in Figure 1.2 using an equation, which can easily be achieved using curve-fitting software. If we do this, we find that our curve of interest has the form,

$$y = -4.9x^2 \quad \text{or} \quad s = -4.9t^2$$

We now differentiate this equation using the rules given in Appendix A:

$$\frac{ds}{dt} = 2 \times (-4.9) \times 0.5^{(2-1)} = -4.9\,\text{m/s}$$

Although there is only a slight difference when compared to the value obtained by the graphical method $(-5.0\,\text{m/s})$, the value found by differentiation is more accurate.

The fact that the curvature of Figure 1.2 is constant is significant — it means that the velocity of the apple changes by a constant amount as it falls. We can prove this by calculating the velocity for a variety of time points and plotting the data as a graph (Figure 1.3) which will have a constant gradient. If we think about what we have done, we see that we have essentially taken the gradient of a gradient, and in so doing have calculated the change in velocity with time. This is known as **acceleration** a, and is represented by

$$a = \frac{dv}{dt} \quad \text{or} \quad a = \frac{d^2s}{dt^2} \tag{1.2}$$

The expression on the right is known as a second derivative and is read as "dee two s by dee t squared." In practice, this means that we perform the

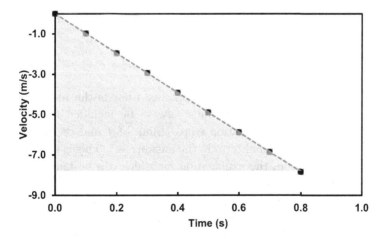

Figure 1.3 Velocity versus time graph for an apple in free fall. The gradient of the graph is equal to the acceleration of the apple and the area under the curve gives the displacement.

differentiation twice:

$$s = (-4.9t^2) \quad \therefore \frac{ds}{dt} = -9.8t \quad \text{and} \quad s = (-9.8t) \therefore \frac{ds}{dt} = 9.8 \text{ m/s}^2$$

$$\therefore a = \frac{d^2s}{dt^2} = 9.8 \text{ m/s}^2$$

Thus, the acceleration of the apple falling from the tree is 9.8 m/s^2, which agrees with the graphical result we obtained from Figure 1.4. In fact, most objects in free fall have $a = 9.8 \text{ m/s}^2$, provided that there is minimal air resistance. This value, known as the acceleration due to gravity, was a major outworking of Newton's work, and would inform his later work on the Universal Law of Gravitation.

Now consider the opposite case where we are given a plot of velocity versus time for an apple in free fall and are asked to determine the total displacement of the apple. This time, instead of calculating the gradient (which would be acceleration as we have already seen) we would calculate the area under the curve. For Figure 1.3, this is very straightforward as the area under the curve is a right-angled triangle:

$$\text{Area} = \frac{\text{Base} \times \text{Height}}{2} = \frac{0.8 \times (-0.78)}{2} = -0.31 \text{ m}$$

Of course, we know the real value of the displacement from our earlier work (-0.39 m) which implies that our calculation above is slightly inaccurate. Instead, it is much better to use an expression for the velocity of the apple and use **integration**. Using the rules in Appendix A, the integral of would be

$$\int -9.8t \ dt = \frac{-9.8t^2}{2} = -4.9t^2 + C$$

There are two things to note in this process. First is the inclusion of dt at the end of the integral. This must always be included as it acts as a boundary where the integration stops (think of $\int$ and dt like a set of brackets). Second is the final term in the answer, $+C$. This is known as the constant of integration, the explanation for which can be found online or in mathematics textbooks.

To remove the constant of integration, we impose limits which define the area we are interested in, i.e., the displacement corresponding to a change in time from t_0 to t_1:

$$\int_{t_0}^{t_1} -9.8t \ dt = \left[\frac{-9.8t^2}{2} + C \right]_{t_0}^{t_1}$$

As we know the values of the limits, $t_0 = 0\,\text{s}$ and $t_1 = 0.8\,\text{s}$, we can substitute these values directly into the expression in square brackets. By subtracting the two expressions, the constants of integration are cancelled and we are left with the desired result.

$$\int_{t_0}^{t_1} -9.8t \ \mathrm{d}t = \left[\frac{-9.8 \times 0.8^2}{2} + C\right] - \left[\frac{-9.8 \times 0^2}{2} + C\right] = -3.1 \ \text{m}$$

Hence, we have shown that integration can be used to work backward from the rate of change to the original data. The applications of integration are much broader than has been shown here. In later chapters, we will use integration to solve more complex differential equations associated with the movement of electrons and radioactive decay.

1.1.2 Newton's second law

Key Point: Any object with mass and velocity is said to have momentum and the rate of change in momentum is known as force.

The equations presented so far have neglected an important physical property of an object — its mass, m. Newton incorporated this by multiplying the velocity of an object by its mass to give the third important parameter in kinematics, **linear momentum**, p:

$$p = mv \tag{1.3}$$

Like velocity, linear momentum is a vector quantity. The rate of change of the momentum of an object is described as **force**, F, and is given by the differential:

$$F = \frac{\mathrm{d}p}{\mathrm{d}t} \tag{1.4}$$

Assuming that mass remains constant, we can see that force is alternatively given by

$$F = m\frac{\mathrm{d}v}{\mathrm{d}t} = ma \tag{1.5}$$

The SI unit of force is the newton, N. One newton can be thought of as the force of the Earth's gravity on a mass of *ca.* 102 g or as a mass of 1 kg having a downward force of 9.81 N.

1.1.3 Newton's third law

Key Point: When two or more bodies interact, their individual momenta act in opposite directions, ensuring that the total momentum is constant.

The last of Newton's laws of motion relates to an important concept in physics, that of the **law of the conservation of momentum**. Think about a bullet being fired from a gun. As the bullet is passing down the barrel of the gun, it experiences a frictional force equal in magnitude to its forward momentum, but the frictional force is acting in the opposite direction. Therefore, the forward momentum must be equal to the backward momentum:

mass of bullet × muzzle velocity = mass of gun × recoil velocity

Essentially, this implies that when two or more bodies act upon one another, the total momentum remains constant, assuming no external forces are also acting.

1.2 Work and Energy

Key Point: The ability of a force to bring about a change in displacement can be used to define the energy of an object.

When a force brings about a change in the displacement of an object, it is said to have done **work**, w (it accelerates, doing positive work):

$$w = Fs \tag{1.6}$$

The work done on the object is measured by its **kinetic energy**, K, which is defined by

$$K = \frac{1}{2}mv^2 \tag{1.7}$$

The SI unit of kinetic energy is the joule, J, which is the same as the SI unit of work. Energy which arises from the position or arrangement of a body is known as **potential energy**, U. If an object of mass m is held stationary at a height h above the ground, the work done when the object falls to the ground would be

$$w = mah \tag{1.8}$$

If the object is close to the Earth's surface, a would be the acceleration due to gravity, g, and the potential energy of the object would be given by

$$U = mgh \tag{1.9}$$

The **law of the conservation of energy** states that the total energy in a system is constant although energy may change from one form to another. For example, suppose there is a 5 kg ball suspended 2 m from the ground. At its current position, its kinetic energy will be zero but its potential energy will be

$$U = 5 \times 9.8 \times 2 = 98 \text{ J}$$

According to the law of the conservation of energy, if the ball is dropped, the potential energy will equal the kinetic energy, i.e.,

$$mgh = \frac{1}{2}mv^2$$

This can be used to calculate the velocity of the ball just before it hits the ground:

$$v = \sqrt{2gh} = \sqrt{2 \times 9.8 \times 2} = 6.3 \text{ m/s}$$

The kinetic energy on impact is therefore given by

$$K = \frac{1}{2}mv^2 = \frac{1}{2} \times 5 \times 6.3^2 \approx 98 \text{ J}$$

Hence, we have shown that the starting energy is equal to the final energy and that the law of the conservation of energy has been obeyed.

1.3 Uniform Circular Motion

Key Point: The rate of uniform circular motion is known as the angular velocity which gives rise to angular momentum.

In the previous section, we saw how Newton's laws could be used to describe the motion of an object along a straight-line trajectory. Of course, objects do not always travel in straight lines. There are numerous examples of an object moving in a curve about some fixed point — the orbit of a Moon around a planet, for example. This introduces the need for laws describing circular motion.

The concept of uniform circular motion is based on the idea of the unit circle. We know that the circumference of a circle is related to its diameter by pi, π (Figure 1.4). It follows that the radius of a circle is related to its circumference:

$$C = 2\pi r \tag{1.10}$$

The angle around the central point of the circle is 360°, but in circle geometry, it is customary to measure angles in **radians**. A radian is the

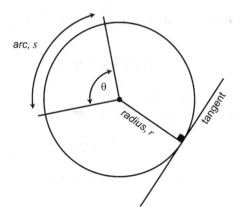

Figure 1.4 Circle geometry. A tangent always forms a perpendicular to the circle (forms a 90° angle).

angle created when the length of the arc is equal to the radius:

$$\theta = \frac{s}{r} \tag{1.11}$$

Therefore, if 360° is equivalent to moving around a full circumference of $2\pi r$, then when moving around a circumference equal to the radius, we get 2π radians:

$$r = 2\pi r \Rightarrow \frac{2\pi \cancel{r}}{\cancel{r}} = 2\pi \quad \therefore \; 360° = 2\pi \, \text{radians}$$

Now consider the movement of a body in a circular path, such as the wheel of a car (Figure 1.5). The rate of change in the angle is defined as the instantaneous **angular velocity**, ω (omega):

$$\omega = \frac{d\theta}{dt} \tag{1.12}$$

If the angle is measured in radians and time is in seconds, ω has units of rad/s. One complete revolution of the car wheel would be a distance of 2π radians (360°). The time it takes to move around the full circumference of a circle is known as the **period**, T, given by

$$\omega = \frac{d\theta}{dt} = \frac{2\pi}{T} \quad \therefore \; T = \frac{2\pi}{\omega} \tag{1.13}$$

We know that cars travel in straight lines; this arises from the tangential velocity created by the rotation of the car wheel. To relate the angular velocity with the tangential velocity, we first imagine that the car wheel

Figure 1.5 Tangential velocity in a car. The angular velocity ω, produces a tangential velocity v_T, in the same direction as the rotation of the wheel.

moves through an arc equal in length to its radius, so that the angle is given by s/r. If we want the rate at which the wheel moves through this arc, we must differentiate with respect to time (keeping r constant):

$$s = \theta r \therefore \frac{\mathrm{d}s}{\mathrm{d}t} = \frac{\mathrm{d}\theta}{\mathrm{d}t} r$$

This result is very similar to Eq. (1.12) which means we can simplify by substituting $\mathrm{d}\theta/\mathrm{d}t = \omega$:

$$v_{\mathrm{T}} = \frac{\mathrm{d}s}{\mathrm{d}t} = \omega r \tag{1.14}$$

It follows that the greater the angular velocity, the greater the tangential velocity (the more the car wheel revolves, the faster the car will move). Finally, we can relate the mass of an object in circular motion to its velocity using the **angular momentum**, L, which usually has units of $\mathrm{kg \cdot m^2/s}$:

$$L = mvr \tag{1.15}$$

1.4 Simple Harmonic Motion

Key Point: Objects in uniform circular motion exhibit a regular periodic displacement on Cartesian axes known as simple harmonic motion. This can be related to the shape of a wave.

When viewed on a displacement–time graph, the movement of a body in uniform circular motion has the form of a **sine wave** in what is known as simple harmonic motion (Figure 1.6). Each part of the sine wave

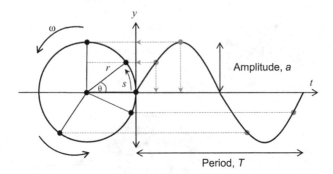

Figure 1.6 General form of a sine wave. The amplitude, a, is the displacement from the x-axis; the period, T, is the time taken for one complete oscillation; the frequency ν, is the number of oscillations per second, which is related to the angular velocity of the body in uniform circular motion.

corresponds to a position on the circumference of a circle. The **amplitude**, a, is the displacement from the x-axis and the **frequency**, ν, is the number of oscillations per second, which is related to the angular velocity of the body in uniform circular motion:

$$\nu = \frac{1}{T} = \frac{\omega}{2\pi} \tag{1.16}$$

To see how the sine function is related to a circle, we can use the construction shown in Figure 1.7. The point A on the unit circle's circumference is found at a distance of one from the x-axis and one from the y-axis (because the radius is one). These two lengths form two sides of a right-angled triangle — the x-length as the adjacent side and the y-length as the opposite side. According to Pythagoras' theorem, the square of the hypotenuse is equal to the sum of the squares of the other two sides. In our case, this would be

$$1^2 = x^2 + y^2 \quad \text{but,} \ 1^2 = 1 \quad \therefore \ 1 = x^2 + y^2$$

In other words, the hypotenuse of our unit circle is one. The ratio of the opposite or adjacent sides to the hypotenuse is related to the angle θ by the sine and cosine functions. In the case of the unit circle, this would be

$$\sin \theta = \frac{y}{x} \quad \text{but,} \ x = 1 \ \text{for unit circle} \ \therefore \ \sin \theta = \frac{y}{1} = y$$
$$\cos \theta = \frac{x}{y} \quad \text{but,} \ y = 1 \ \text{for unit circle} \ \therefore \ \cos \theta = \frac{x}{1} = x$$

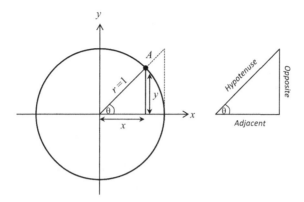

Figure 1.7 Trigonometric identities in a unit circle. The hypotenuse is a distance of one unit from the origin. Therefore, by Pythagoras' theorem, the opposite and adjacent sides must also be a distance of one unit from the origin.

It follows that any point on the circumference of a circle of radius r will have xy coordinates given by the parametric equations of a circle:

$$\sin\theta = \frac{y}{r} \quad \therefore \quad y = r\sin\theta \quad \text{and} \quad \cos\theta = \frac{x}{r} \quad \therefore \quad x = r\cos\theta$$

The angular velocity of a body in uniform circular motion can be given by Eq. (1.12); rearrangement of this and substitution into either of the parametric equations of a circle give the basic equation for a sine wave or cosine wave:

$$y = r\sin\omega t \tag{1.17}$$

$$x = r\cos\omega t \tag{1.18}$$

In fact, a cosine wave can be described as being *sinusoidal* because sine and cosine are related.

Looking at the two sine waves in Figure 1.8, we see that wave A starts at $x = 0$, but wave B starts at some other value to the left of the origin. This difference is due to the *phase*, φ (phi), of the waves which is defined as the x-value when the wave starts its oscillation. The *phase difference* in Figure 1.8 is 45° (or $\pi/2$ radians). The phase can be included in the equation for a sine or cosine wave, e.g.,

$$y = r\sin(\omega t + \varphi) \tag{1.19}$$

Finally, we can relate the period of the wave to the equation for a sine wave by expanding the ω term:

$$y = r\sin\frac{2\pi}{T}t \tag{1.20}$$

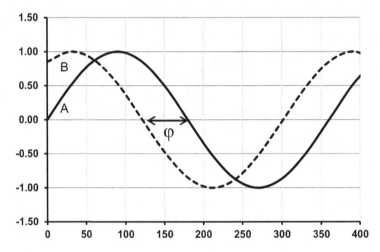

Figure 1.8 Phase of sine waves. Wave A starts at $x = 0$ while wave B starts at $x > 0$ which creates a phase difference.

Another useful example of simple harmonic motion involves **Hooke's law**. Imagine there is a mass, m, attached to a spring which is in turn attached to a rigid support. If the mass is displaced from its resting position by a distance x, the spring will exert a restoring elastic force, F. The magnitude of this force is proportional to the displacement, $F \propto x$, and by including an experimentally determined proportionality constant, k, we arrive at

$$F = -kx \qquad (1.21)$$

Recalling Newton's second law of motion, Eq. (1.5), we see that on combining with the expression for Hooke's law, we get

$$ma = -kx$$

To show the acceleration experienced by the mass as it is displaced, we expand the a-term to that given in Eq. (1.2):

$$m\frac{\mathrm{d}^2x}{\mathrm{d}t^2} = -kx$$

If we rearrange this equation and make the substitution $\omega^2 = k/m$, we arrive at the equation for the **harmonic oscillator**:

$$\frac{\mathrm{d}^2x}{\mathrm{d}t^2} = -\omega^2 x \qquad (1.22)$$

Assuming that Eq. (1.22) describes simple harmonic motion, the solution to this equation should be that of a sine wave, i.e., Eq. (1.18). We can demonstrate this by differentiating Eq. (1.18) twice:

$$\frac{\mathrm{d}x}{\mathrm{d}t} = -r\omega \sin \omega t \quad \text{and} \quad \frac{\mathrm{d}^2 x}{\mathrm{d}t^2} = -r\omega^2 \sin \omega t$$

Noting that $x = r \sin \omega t$ in the original version of Eq. (1.18) we see that Eq. (1.22) is obtained. Alternatively, Eq. (1.22) can be integrated (this is shown in Appendix A).

1.5 Waves

Key Point: A wave is a periodic disturbance which carries energy at definite speed; many naturally occurring waves are sinusoidal.

We are familiar with the idea of a wave in the general sense that it involves a periodic movement — the tides at the seaside move toward the beach at regular intervals. In this case, we are speaking of a ***mechanical wave***, i.e., a wave which travels through a medium. More precisely, we say that a mechanical wave is a disturbance that travels through a medium with finite velocity and that it transfers energy from one point to another. We can classify mechanical waves into three categories:

(1) Transverse waves, such as seismic waves.
(2) Longitudinal waves, such as sound waves.
(3) Surface waves, such as waves in a pool of water.

Waves which do not require a medium to travel through, although can travel through matter to various extents, are known as ***electromagnetic waves***. In some respects, these waves have some of the properties of transverse waves, except that they can travel in vacuum and have a more complex shape. As the name implies, electromagnetic waves have two components: an electric field, E, and a magnetic field, B; these two components can be represented as two sinusoidal waves which oscillate at right angles to each other. Electromagnetic waves are produced by charged particles (mostly electrons) and they travel in vacuum at the speed of light. For reasons which will be discussed in Chapter 2, the energy of electromagnetic waves is contained in discrete packets known as ***photons***.

Waves now have a familiar form and are fundamentally described by their ***wavelength***, λ (lambda), and frequency. To incorporate the speed of a wave, we imagine that a wave is travelling along the positive x-direction with constant speed, v, which for an electromagnetic wave will be the speed of light. As speed is the change in displacement over change in time and the period of a wave is the time taken for one wavelength to pass a point, we can express speed as a function of wavelength and frequency:

$$v = c = \frac{s}{t} \approx \frac{\lambda}{T} \text{ but since, } T = \frac{1}{\nu} \text{ we have, } c = \frac{\lambda}{\left(\frac{1}{\nu}\right)}$$

$$\therefore c = \nu \times \lambda \tag{1.23}$$

If we want to describe the shape of a wave, we first express its sine function in terms of its wavelength:

$$y = a \sin \frac{2\pi}{\lambda} x \tag{1.24}$$

We now replace y with a new entity describing the disturbance caused by the wave; this is known as the **waveform**, ψ (psi). For a sound wave, the waveform would correspond to the displacement of air by the wave and for an electromagnetic wave, it could correspond to the value of the electric field. As the waveform is a function both of position (x) and time (t), it is usually written as $\psi(x,t)$. The movement of the wave along the positive x-direction is given by

$$\psi(x,t) = a \sin \left[\frac{2\pi}{\lambda}(x - vt) \right] \tag{1.25}$$

This is the equation for a **travelling wave** which is a useful starting point for discussion on wave mechanics (Chapter 2).

Now, consider a wave formed by a string attached to a rigid support at both ends which are a distance L apart (Figure 1.9). As the string is attached at both ends, its displacement must be zero at these points. If we label the first position as $x = 0$ and the second position as $x = L$, we can write the **boundary conditions** for the wave:

$$\psi(0,t) = 0 \quad \text{and} \quad \psi(L,t) = 0$$

Such a wave is known as a **standing wave** and the boundary conditions are physical constraints placed on the wave. These boundary conditions are satisfied by a sinusoidal waveform. First, we tidy up Eq. (1.25) by expressing v in terms of λ and T:

$$\psi(x,t) = a \sin 2\pi \left(\frac{x}{\lambda} - \frac{t}{T} \right)$$

Now, introducing a new relation known as the **wavenumber**, k, and substituting this into the previous expression, we obtain

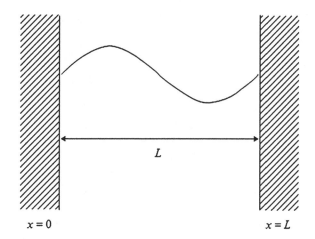

Figure 1.9 A standing wave with boundary conditions at $x = 0$ and $x = L$.

Let,

$$k = \frac{2\pi}{\lambda} \quad \therefore \quad \psi(x,t) = a \sin kx - 2\pi \left(\frac{t}{T} \right)$$

Finally, substituting Eq. (1.16) into our last expression, we obtain the desired result:

$$\psi(x,t) = a \sin(kx - \omega t) \tag{1.26}$$

When a progressive wave is incident on a boundary, it is reflected back along the same path and the resultant disturbance will be the sum of the two waveforms:

$$\psi_{\text{forward}} + \psi_{\text{reverse}} = a \sin(kx - \omega t) + a \sin(kx + \omega t)$$

This is solved using the trigonometric relationship given in Eq. (1.27) which gives our final result, Eq. (1.28):

$$\sin A + \sin B = 2 \sin \frac{A+B}{2} \cdot \cos \frac{A-B}{2} \tag{1.27}$$

$$\therefore \quad \psi(x,t) = 2a \sin kx \cdot \cos \omega t \tag{1.28}$$

This equation no longer represents a progressive wave because x and t do not occur together in the equation. It describes harmonic motion ($\cos \omega t$) with an amplitude ($2a \sin kx$) that varies with position along the string. Equation (1.28) clearly satisfies the first boundary condition ($x = 0$) because $\sin(0) = 0$. The second boundary condition ($x = L$) is also satisfied

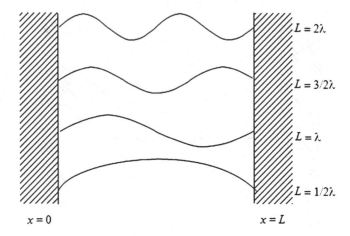

$x = 0$ $\qquad\qquad\qquad\qquad\qquad\qquad x = L$

Figure 1.10 A series of standing waves with $n = 1, 2, 3$, and 4 and boundary conditions $x = 0$ and $x = L$.

provided that $\sin kL = 0$. It is only possible for $\sin kL = 0$ for specific values of kL which are positive integer multiples of π:

$$kL = n\pi \quad n = 1, 2, 3, 4, \ldots$$

Recalling that k is a function of wavelength, we can express this boundary condition as

$$L = n \cdot \frac{1}{2}\lambda$$

We have therefore shown that the boundary conditions of a standing wave are satisfied by a sinusoidal waveform.

A series of standing waves are shown in Figure 1.10 for $n = 1, 2, 3$, and 4. We see that for $n = 1$, the amplitude of the wave is only zero at the boundaries and is maximal half-way between the boundaries. The point at which the amplitude is minimal is referred to as a **node**; similarly, the point where the amplitude is maximal is known as an **anti-node**. We see that for a standing wave, nodes occur at multiples of half-wavelength. The existence of nodes and anti-nodes are an important consideration in one interpretation of quantum mechanics as they are related to the energy of an electron in a particular quantum state.

1.6 Relativity

Key Point: The laws of physics are constant within a specified region, but they may appear to be violated from an outside perspective.

The scientific revolution that Newton started with the *Principia* in 1687 was more or less completed by his ideas on **relativity**. At the time, a major question in science related to the rotation of the Earth. Galileo Galilei (1564–1642) published a work in which he described a thought experiment involving the motion of a ship:

(1) Imagine you are in a windowless cabin on a ship which is moving at constant velocity on calm seas. To you, it would feel as if the ship were motionless, despite the fact that it is moving.

(2) If you drop a ball in your cabin, it will fall along a straight path to the floor of the cabin, just as it would if the ship were stationary.

(3) If the experiment was repeated, but this time it was being watched by someone outside the ship, they would see the ball move vertically (as you do), but also horizontally due to the motion of the ship.

From your point of view, the ball has no horizontal velocity, but for outside observers, it would have horizontal velocity. In other words, velocity depends on the **reference frame** in which it is measured. Galileo used this to explain that the Earth could be rotating at a constant velocity without affecting how objects behaved on or near the surface. Nowadays, this concept is known as Galilean relativity and Newton applied his laws of motion to Galilean relativity to create Newtonian relativity. Two key features of Newtonian relativity are:

(1) Velocity is not absolute; it depends on the reference frame.

(2) Time is an absolute constant.

Newtonian relativity remained the accepted theory until the late 1800s when it became clear that it could not explain experimental observations relating to light.

Early experiments on the nature of light revealed wave-like properties.[a] In the 1800s, James Clerk Maxwell calculated the speed of light in vacuum, c, as 2.99×10^8 m/s, but this presented a problem as Newtonian relativity states that the speed of light must be relative to something. Since light was considered as a wave, and waves need a medium to travel through, it was believed that light travels through a "luminiferous ether." In 1887, Albert Michelson (1852–1931) and Edward Morley (1838–1923) sought to determine the velocity of the Earth's rotation in the luminiferous ether by measuring the speed of light at two different reference frames. By subtracting the differences in the speed of light, they hoped to show the

---

[a]For example, Thomas Young's (1773–1829) double slit experiment which showed that light created interference patterns which could only be explained by wave-like behavior.

absolute speed of the Earth in space. When they calculated their result, however, they found that the value obtained was zero, which was clearly wrong because astronomical observations had proved that the Earth rotates about its axes. This was the first failure of Newtonian physics and left many physicists uncertain about how to explain these observed phenomena.

A major paradigm shift occurred in 1905 when Albert Einstein (1879–1955) published his paper "Zur Elektrodynamik bewegter Körper" ("On the Electrodynamics of Moving Bodies"). This work proposed explanations for experimental observations that could not be explained by classical (Newtonian) physics. To start, Einstein postulated the following:

(1) The laws of physics are invariant (they must be the same in all stationary points of view, i.e., within the same reference frame).
(2) The speed of light in vacuum is constant, regardless of any motion.

These postulates formed part of his *special theory of relativity*. In this, Einstein replaced the classical view that space and time were separate entities with the idea that they were interconnected as a single continuum known as spacetime.[b] In special relativity, spacetime is considered to have four dimensions, that is, the three dimensions of space (x, y, z) and a fourth dimension, time. This is known as Minkowski spacetime.[c]

Of the various outworkings of special relativity, the one which concerns us most is the famous mass–energy equivalency, instantly recognizable as

$$E = mc^2 \qquad (1.29)$$

To illustrate the meaning of Eq. (1.29), think about the following: suppose we have two particles which have mass but negligible energy or momentum. When these two particles collide, they annihilate one another, releasing energy in the form of electromagnetic radiation which travels at the speed of light. In other words, there is a 100% conversion of mass to energy.[d] Einstein stated that even when an object is at rest, it will still possess some energy, known as its *rest energy*, which is always associated with

---

[b]The idea of spacetime was pioneered by Hendrik Lorentz who created a mathematical framework (Lorentz transformations) to show how the speed of light could be independent of reference frames.
[c]Einstein published a later theory, general relativity, which incorporated gravity. In this, spacetime is curved due to the presence of matter, which is predicted by the Einstein field equations.
[d]In this example, we are speaking of the annihilation of an electron and a positron which produces two gamma ray photons.

its **rest mass**. Of course, this means that mass can be converted to energy and *vice versa*. For example, 1 kg is equivalent to 9×10^{16} J:

$$E = mc^2 = 1.0 \times (2.99 \times 10^8)^2 \approx 9 \times 10^{16} \text{ J}$$

During a chemical reaction, such as neutralization which is exothermic, the conversion of mass to energy is almost negligible. However, in nuclear fission, where a heavy element is split into two fragments, the change in mass is small, but the energy released is large. This is of vital importance to the nuclear power industry.

The final important implication of special relativity is that if an object with mass were to accelerate to close to the speed of light, its **relativistic mass** would increase. The relativistic mass is the mass that would be observed from a different reference frame to that occupied by the object. In other words, if you and the object were both travelling at the speed of light (i.e., in the same reference frame) the object would appear to have its ordinary rest mass. It is only to an outside observer in a different reference frame that the mass would seem to increase. In chemistry, a good example of this is an electron orbiting close to the nucleus of a heavy element such as mercury. The relativistic mass of a 1s electron in mercury is about 1.23 times its rest mass, which causes a significant contraction of the 1s orbital. This causes the outer 6s electrons to be held very tightly to the atom, essentially preventing it from forming a bond with another mercury atom. This is why mercury is a liquid at room temperature.

1.7 Electrostatics

Key Point: Coulomb's law describes the force which arises between two point charges; this follows an inverse square law.

Since antiquity, it has been known that certain materials have the ability to build up an attractive or repulsive force — for example, rubbing amber rods with animal fur. We now recognize this is being due to the formation of an electrical charge, created by an uneven distribution of electrons. We will see in Chapter 2 that a major task in early atomic physics was to establish the value of the charge on an electron. In coulombs, this is

$$e = 1.602 \times 10^{-19} \text{ C}$$

This value is often referred to as the **elementary charge** and is used to express the charge of other particles, such as protons, in relative terms.

The field of electrostatics concerns the behavior of charged particles and is largely dominated by the work of Charles de Coulomb (1736–1806). Early experiments showed that when two identical charges, q_1 and q_2, were in close proximity, they experienced a repulsive force proportional to the size of the charge and inversely proportional to the square of the distance separating the charges.

$$F \propto q_1 \quad F \propto q_2 \quad F \propto \frac{1}{d^2}$$

We can combine these expressions with a constant of proportionality to give **Coulomb's law**:

$$F = k_e \left(\frac{q_1 q_2}{d^2} \right) \tag{1.30}$$

The charge is taken in coulombs, distance in meters, and force in newtons. The constant, k_e, has the value $8.98755 \times 10^9 \, \text{N.m}^2/\text{C}^2$ which is often expressed in terms of a more fundamental constant, the **permittivity of free space**, ε_0,

$$k_e = \frac{1}{4\pi\varepsilon_0} \tag{1.31}$$

where $\varepsilon_0 = 8.85419 \times 10^{-12} \, \text{C}^2/\text{N.m}^2$; this describes the resistance encountered by an electric field in vacuum. Therefore, Coulomb's law may be written in the equivalent form:

$$F = \frac{q_1 q_2}{4\pi\varepsilon_0 d^2} \tag{1.32}$$

This version of Coulomb's law appears quite often in chemistry; for example, it is used to explain certain features of atomic theory and the separation of ions in a mass spectrometer.

1.8 Emergence of Quantum Physics

Key Point: Quantum mechanics explains physical properties on the microscopic scale.

In the previous sections, we have explained in some detail the foundations of classical physics. If we were to try and summarize classical physics, we could possibly arrive at the following statements:

(1) An object travels along a trajectory which allows its precise position and momentum to be described at any instant.
(2) Any object in motion obeys Newton's second law ($F = ma$); it can be accelerated through gain of extra energy.

(3) Waves and particles are distinct entities described by separate equations.

There is a common misconception that modern physics has disproved the work of Newton *et al.* This is not the case — the kinetic energy of a falling object still obeys Newton's second law. However, in the interceding years, a more fundamental set of laws have arisen, which on the macroscopic scale are indistinguishable from those of Newton. These laws, referred to as quantum mechanics, explain experimental observations at the microscopic scale which could not be explained by the classical approach.

In Chapter 2, we will see how the failure of classical physics leads several key scientists to formulate the laws of quantum mechanics. Central to this theory is the idea that the physical properties of a system are constrained to take on specific values, known as quanta, and that matter may possess both the properties of a particle and a wave. This latter feature introduces the idea of uncertainty, for when the velocity of a microscopic particle is known, its position cannot be accurately determined.

1.9 Basics of Atomic Structure and Radioactivity

Key Point: Unstable atoms gain stability by releasing nuclear mass as ionizing radiation.

The chemistry of an atom is dictated by the number and arrangement of its electrons, with the nucleus providing a positive charge to bind the electrons in atoms and molecules. The protons and neutrons found in the nucleus are held together by three forces which govern its stability. Some nuclei are unstable and undergo radioactive decay to gain stability. For example, the nucleus of an atom of uranium undergoes radioactive decay and releases an *alpha particle*, forming a new element, thorium:

$$^{238}_{92}\text{U} \rightarrow {}^{234}_{90}\text{Th} + \alpha$$

You will see later than an α-particle is really the nucleus of a helium atom. The thorium atom produced is also unstable and undergoes further radioactive decay releasing another form of radiation known as a *beta particle* and producing another new element, protactinium:

$$^{234}_{90}\text{Th} \rightarrow {}^{234}_{91}\text{Pa} + \beta$$

For reasons that will be discussed later in Chapter 4, when a nucleus releases an alpha or beta particle, excess energy is released in the form of a third

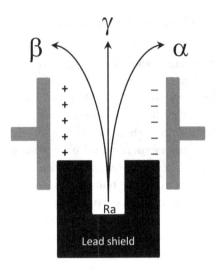

Figure 1.11 Effect of an electric field on radiation. Alpha particles are attracted toward the anode; beta particles are attracted toward the cathode; and gamma rays remain unaffected suggesting that they are uncharged.

type of radiation, known as a **gamma ray**. We can rewrite the decay of uranium to reflect this:

$$^{238}_{92}\text{U} \rightarrow {}^{234}_{90}\text{Th} + \alpha + 2\gamma$$

The three types of ionizing radiation can be separated on the basis of their physical properties (Figure 1.11). Alpha particles have been shown to be relatively slow moving with a small mass-to-charge ratio. Conversely, beta particles are fast moving with a large mass-to-charge ratio. Gamma rays are unaffected by a magnetic field and thus are shown to be a form of electromagnetic radiation, similar to light, but with a much shorter wavelength. Due to their different physical properties, the different forms of radiation are said to have different penetrating power. Alpha particles are stopped by thin paper or human skin, but beta particles are only stopped by a metal plate about 5 mm thick. Gamma rays can penetrate several centimeters of lead, but are stopped by thick concrete.

1.10 Particle Physics

Key Point: Fundamental particles are those which are not composed of any smaller entity; electrons are fundamental particles, but protons and neutrons are not.

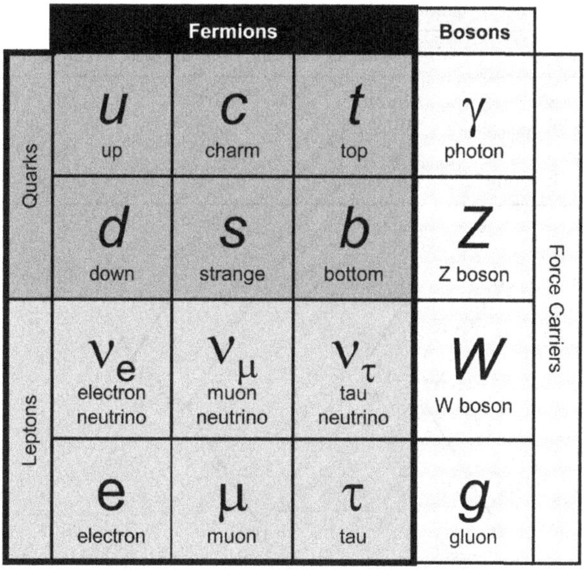

Figure 1.12 The standard model. The quarks and leptons are regarded as fundamental particles, and are collectively known as fermions as they have half-integer spin. The bosons shown are responsible for three of the fundamental forces (gravitons are omitted).

Since the 1940s, experiments with high-energy particle accelerators have revealed hundreds of unstable particles with exceedingly short half-lives. From this, it is now in our understanding that all matter is composed of elementary particles known as *quarks* and *leptons* which impart the key properties of mass, charge, and spin to other larger, composite particles. There are six types of quarks which combine in different ways to produce the *hadrons* which are further subdivided into the mesons and baryons. The leptons are believed to have no internal structure and are grouped into three families, including the electron and electron neutrino family which is important in radioactive decay (Figure 1.12).

A key aspect of particle physics is the existence of four fundamental forces in the universe which interact with particles through field-carrying exchange particles known as *gauge bosons* (Table 1.1). The *gravitational force* has the longest range and is responsible for holding stars, planets, and galaxies. The gravitational field is created by the exchange of gravitons. Although these have never been observed, their existence is supported by different theories arising from particle physics. The *electromagnetic force* is responsible for the binding of electrons and protons in atoms and

Table 1.1 The Fundamental Forces

Force	Mediators	Relative Strength	Range (m)
Strong	Gluons	10^{38}	10^{-15}
Electromagnetic	Photons	10^{36}	∞
Weak	Bosons	10^{25}	10^{-18}
Gravitational	Gravitons	1	∞

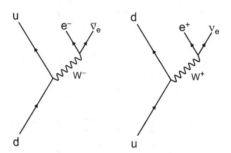

Figure 1.13 Feynman diagrams. The diagram on the left-hand side shows beta decay where a down quark is transformed to an up quark. The reverse process, shown on the right-hand side, produces a positron and is part of the fusion process in the Earth's Sun.

molecules to form matter. It is mediated by the exchange of virtual photons between particles.

The **weak interaction** arises between fermions through the exchange of W^+, W^-, and Z bosons. The weak nuclear force is particularly relevant to our study of nuclear chemistry as it is responsible for beta decay and the nuclear reactions which drive energy production in the Sun. In the former case, a neutron is converted to a proton when a down quark releases a W^- boson, forming an up quark. The W^- boson rapidly decays to give an electron (the beta particle) and an anti-neutrino. This process can be conveniently represented using a **Feynman diagram** (Figure 1.13), which is a pictorial representation of the particle exchange process. Particles are represented by straight lines with arrows which represent the direction of their travel. Virtual particles are depicted as wavy or broken lines without arrowheads. The vertical direction gives a nominal indication of time. We can similarly represent the transformation of a proton into a neutron, releasing a positron and neutrino, as it occurs in the hydrogen fusion reaction in the Sun.

The **strong interaction** is responsible for the formation of hadrons, such as the protons and neutrons, and is responsible for much of the mass of these particles (due to mass–energy equivalency). When the strong interaction has acted to bind together quarks, the residual energy is used to bind together protons and neutrons to form a nucleus. This residual energy is known as the nuclear force and is in turn mediated by exchange particles called mesons.

The formation of quarks about 10^{-30} s after the Big Bang marked the beginning of the formation of nuclei. However, this was not a rapid process. Current theories of cosmology state that the universe formed from a dense, point-like singularity roughly 14 billion years ago. Such was the intensity of the Big Bang, it is believed that all four fundamental forces existed as a unified force. Then, *ca.* 10^{-35} s after the Big Bang, the gravitational force separated from the remaining forces and marked the formation of quarks and leptons. At this stage, the strong interaction became a separate identity, leaving the remaining two forces (collectively known as the electroweak force). It was not until the universe cooled to below 10^{15} K that protons and neutrons condensed out of the quark–gluon plasma. Accompanying this, the electroweak force separated into the electromagnetic and weak interaction, which marked the formation of the first nuclei which later combined to form the first atoms.

Chapter Summary

- Classical physics is dominated by rules based on Newton's laws of motion. These link the kinematic properties of a body in motion with mass and force.
- Energy in classical physics is initially divided into kinetic or potential energy and is described as a function of mass and velocity.
- Newtonian relativity introduces the idea of reference frames and explains how the laws of physics may appear to be violated from outside the reference frame. Special relativity introduced the idea that time is a universal constant and that mass and energy are equivalent.
- Simple harmonic motion, which is based on uniform circular motion, provides a mathematical description of the behavior of waves and other systems. Electromagnetic radiation has a sinusoidal waveform and travels through a vacuum at the speed of light.

- Quantum physics explains the behavior of microscopic systems and is based on the idea that the properties of microscopic systems are constrained to specific values.
- Unstable atoms gain stability through radioactive decay, which eventually converts one element into a different element. This is the basis of nucleosynthesis.
- Subatomic particles are composed of smaller entities known as quarks. These are described by the Standard Model.

Review Questions

(1) A particle moving in a straight line has velocity given by the equation $v = 2t^2 - 7t + 3$. Find (i) the velocity at $t = 0$; (ii) the time when $a = 0$; (iii) the displacement at $t = 1/2$.

(2) A motorbike increases its speed from 0 to $5.2 \, \text{m·s}^{-1}$ in 0.832 s. If the motorcyclist weighs 70 kg, what is (i) the linear momentum of the system; (ii) the average force experienced by the motorcyclist?

(3) A 900 kg car is travelling at 60 miles per hour. Calculate its kinetic energy at (i) this speed; (ii) at 30 mph; (iii) what is the relationship between speed and kinetic energy?

(4) Calculate the angular momentum for the Earth's Moon in December, taking $r = 3.84 \times 10^8 \, \text{m}$ and $m = 7.35 \times 10^{22} \, \text{kg}$.

(5) A travelling sinusoidal wave has a speed of 5 m/s with a wavelength of 4 m and amplitude of 0.05 m. Evaluate the waveform of the wave at $x = 6 \, \text{m}$ and $t = 1 \, \text{s}$.

(6) Calculate the wavelength of a radiowave with a frequency of 3 kHz, taking $c = 2.99 \times 10^8 \, \text{m·s}^{-1}$. You should state your final answer using an appropriate multiple of the SI unit.

(7) Determine the energy released in the reaction $p^+ + p^- \rightarrow 2\gamma$, taking $m = 1.67 \times 10^{-27} \, \text{kg}$ and $c = 2.99 \times 10^8 \, \text{m·s}^{-1}$.

(8) Calculate the force of repulsion between two 1 C charged particles separated by (i) 0.25 m; (ii) 0.5 m; (iii) 1 m. Explain the change in the magnitude of the force.

(9) Provide an equation for the decay of (i) gold-198 to mercury-198; (ii) plutonium-239 to uranium-235; (iii) bismuth-212 to thallium-208.

(10) Suggest why strontium-90, a beta emitter, is used in preference to americium-241, an alpha emitter, in industrial equipment to measure the thickness of aluminum sheet.

Chapter 2

The Structure of the Atom

"Erwin with his psi can do
Calculations quite a few
But one thing has not been seen
Just what does psi really mean?"

W. Hückel

From the perspective of nuclear chemistry, we can explain nuclear and chemical phenomena using comparatively simple models of the atom, many of which are familiar to us from school-level physics. On completion of this chapter and the associated questions you should:

- Be familiar with the historical development of atomic structure.
- Be able to interpret key experimental evidence supporting various models of the atom.
- Appreciate the origins of the modern (Schrödinger) atom and why it was necessary.
- Understand the relationship between the nucleus and the surrounding electrons.

2.1 Brief History of the Atom

Key Point: The idea of the atom as an indivisible entity started with the ancient Greek philosophers and forms the basis of modern atomic theory.

For many people, atoms are regarded as the basic building blocks of matter. This idea stems from antiquity when a number of ancient philosophers proposed the idea of "*atomism*." The key emphasis here is that this is a philosophical interpretation; it was not based on any empirical evidence. Many well-known historical figures were involved in the development of these theories — the Greeks Plato and Aristotle, the Islamic theologian Al-Ghazali, and the Roman physician Galen. The theory of atomism centered on the belief that all matter was composed of tiny, indivisible entities known as atoms. This idea later developed into corpuscularianism, espoused by Robert Boyle in his book *The Sceptical Chymist*, who believed that atoms could be divided into smaller entities. Experimental evidence began to accumulate, and notable chemists such as John Dalton began to develop the atomic theory most recognizable to us today.

2.2 The Thomson Model

Key Point: Thomson discovered the electron and established an early model of the atom known as the Plum Pudding Model.

Sir John J. Thomson (1856–1940) was the Cavendish Professor of Physics at the University of Cambridge between 1884 and 1919 during which time he established himself as a gifted teacher and physicist. In his time at the Cavendish laboratory, Thomson was involved in a number of significant discoveries, including the first proof of the existence of the electron, identifying isotopes of a stable element and developing the first mass spectrometer. However, it was his experiments with cathode ray tubes for which he is most remembered as this provided the first experimental proof for the existence of electrons.

Cathode rays are produced at a negative electrode in an evacuated tube commonly known as a cathode ray tube. Early versions of this apparatus, known as *Crookes tubes*, employed a high electric potential difference between a separated anode and cathode. The cathode rays produced interacted with fluorescent materials to create a green glow. Crookes believed that when a current was applied, "radiant matter" was produced at the cathode which was then attracted toward the positive anode. Thomson partly agreed with this and in a series of ingenious experiments, he sought to prove that Crookes' cathode rays were actually a stream of negatively charged particles.

In the first of his experiments, Thomson noticed that he could deflect the path of the cathode rays with an electric field produced by a pair of electrically charged plates (Figure 2.1). The cathode rays were attracted toward the positively charged plate, proving that they carried a negative

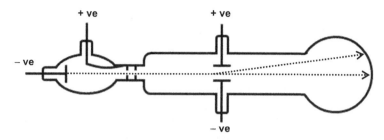

Figure 2.1 Thomson's experiment with a Crookes tube. The cathode ray is deflected toward the positive plate suggesting that cathode rays carry a negative charge.

charge. The next stage was to examine the effect of a magnetic field on the path of the cathode rays. Thomson employed a Helmholtz coil — a pair of coiled wires which produce a uniform magnetic field when a current is supplied — found that when the magnetic field was present, the cathode rays were deflected. The real genius of Thomson's work was to balance the deflection of the cathode rays due to the electric and magnetic fields by adjusting the field strengths. From Maxwell's work, he knew that the ratio of the electric and magnetic fields would give the velocity of the cathode rays. From this, Thomson obtained a value around $2 \times 10^5 \, \mathrm{ms}^{-1}$, which although two orders of magnitude smaller than the current value, was a crucial step toward a quantitative interpretation of the atom.

Thomson felt certain that cathode rays were actually composed of particles which he initially named corpuscles. His next challenge was to determine the specific charge of these corpuscles, which he realized he could do by removing the magnetic field, measuring the deflection of the corpuscles due to the electric field alone and relating this to the acceleration imparted by the electric field. From this, he was able to obtain a value for the ratio of mass to charge (m/z) (the specific charge).

Thomson felt sure that corpuscles represented a smaller unit of matter than the atom. He noted that as the value of the specific charge for these corpuscles was more than two thousand times smaller than that of a hydrogen atom, a hydrogen atom must contain more than two thousand corpuscles. This presented a problem, as having more than two thousand corpuscles would impart a considerable negative charge on the hydrogen atom, which was inconsistent with experimental evidence. Therefore, to balance this, Thomson believed that the negative corpuscles were suspended in a sphere of positive charge to balance the overall charge. This idea, which became known as the ***Plum Pudding Model*** of the atom, was published in 1904 and was generally well accepted by the scientific community. Significantly, it represented the first serious attempt at describing the composition of matter using experimental data.

2.3 The Oil Drop Experiment

Key Point: Millikan characterized the electron and was able to provide a fairly accurate measure of its charge (the fundamental charge).

Thomson's work identified the specific charge of what would later become known as the electron. However, it was the American physicist Robert Millikan (1868–1953) who would make the next significant discovery on the nature of electrons. Working with his colleague Harvey Fletcher (1884–1981), Millikan conceived of an experiment in which the charge of a single electron could be measured and allowed the determination of the mass of an electron using Thomson's value for the specific charge.

In their apparatus, Millikan and Fletcher placed two horizontal metal electrodes at fixed uniform distance apart. When oil droplets were sprayed into an upper chamber they would fall toward the charged plates and acquire a charge. Some of the drops would then rise upwards (due to their charge). If a single drop is selected for observation and the electric field is turned off, this drop falls, allowing its weight to be determined. When the charge was varied, Millikan noted that it was always a multiple of -1.592×10^{-19} C which he proposed was the charge on an electron.

Millikan and Fletcher's work, which became known as the *oil drop experiment* and was published in 1910, produced a value for the charge of an electron in good agreement with the accepted modern-day value $(-1.602 \times 10^{-19}$ C). Millikan attracted criticism over allegations of data manipulation which may have reduced his experiments' standard error from 2 to 0.5%. Regardless of this, however, Millikan and Fletcher were able to define what would turn out to be a fundamental constant in physics and Millikan was awarded the Nobel Prize for Physics in 1923.

2.4 The Rutherford Model

Key Point: Rutherford's interpretation of the gold-leaf experiment lead to the discovery of the nucleus and the proposition of a solar system model of the atom.

Ernest Rutherford (1871–1937) was a New Zealand born physicist who completed his doctorate under the supervision of J.J. Thomson. He made a number of significant discoveries, particularly in the field of radioactivity which lead to him receiving the Noble Prize for Chemistry in 1908. It was after this period of research, however, that Rutherford was to make his most significant discovery — that of the atomic nucleus. Working with Hans Geiger (1882–1945) and Ernest Marsden (1889–1970), Rutherford studied the release of energetic particles from radium.[a] They found that radium released a specific type of particle which could travel only short distances in air and interacted with fluorescent materials to produce flashes of light (known as scintillations). Rutherford and coworkers studied how these particles were absorbed by different materials, including gold leaf. Their experimental set-up was relatively straightforward: a radium source was contained in a lead block with a tunnel exiting at one side, which collimated the beam of particles and directed them toward the target. The trajectory of the particles could be followed using a fluorescent screen (Figure 2.2).

Assuming Thomson's model was correct, the particles should have passed through the gold leaf to produce a distinctive pattern on the fluorescent screen. To Rutherford's surprise, some of the particles showed very large deflection angles and some were even bounced back toward the radium source. Rutherford is quoted as saying "it was quite the most incredible event that has ever happened to me in my life. It was almost

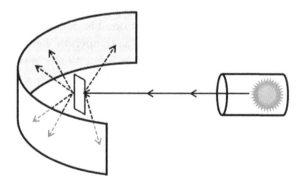

Figure 2.2 Rutherford–Geiger–Marsden gold leaf experiment. Incident alpha particles are deflected in various directions and detected on a fluorescent screen. Backscattering suggested the presence of a dense, central nucleus.

---

[a]These experiments were conducted with radium bromide, probably originally prepared from pitchblende by Pierre and Marie Curie.

as incredible as if you fired a 15 inch shell at a piece of tissue paper and it came back and hit you."

The magnitude of the large deflections could only be explained by the presence of a positive charge localized within a very small volume at the core of the atom. This core, now known as the nucleus, was believed to be surrounded by a cloud of electrons, making the atom electrically neutral. The Rutherford atom, as it became known, was haphazard in the sense that the electrons were just believed to be in a random arrangement outside the core of the atom. Although a few attempts were made to describe arrangement of electrons in an atom, it was the Danish physicist Niels Bohr who made the most credible contribution to this stage of the development of an atomic theory. Bohr's theory relied heavily on the new idea of the quantum, first published in 1901 by Max Planck, and which described the quantization of energy.

2.5 Planck and Einstein — The Quantization of Energy

Key Point: Energy cannot have continuous values but is restricted to multiples of Planck's constant.

Prior to the 1900s, physics was dominated by the classical treatment pioneered by Newton, Boltzmann, and Maxwell. One area of physics where this created a problem was in the modeling of **black body radiation** using the Rayleigh–Jeans law. A black body is an entity that at constant temperature emits electromagnetic radiation. The wavelength of the electromagnetic radiation is dependent upon the temperature of the black body and we see that when radiance is plotted as a function of wavelength for a black body source such as the Earth's Sun ($T \approx 6000\,\mathrm{K}$), a graph similar to Figure 2.3 is obtained. This is clearly significantly different from the observed data which predicts a maximum at around $500\,\mathrm{nm}$. A similar trend is observed using Wien's law which shows that the wavelength of the electromagnetic radiation is inversely proportional to the temperature of the black body.

The discrepancy between the observed and predicted lines on Figure 2.3 can now be explained retrospectively by quantum physics; however, at the time of its discovery, there was no obvious explanation arising from statistical mechanics and this was consequently dubbed the "*UV catastrophe*." The source of the problem was that statistical mechanics depends on laws for the macroscopic scale (relatively speaking), while the fundamental processes responsible for black body radiation are microscopic in nature.

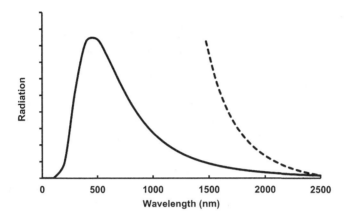

Figure 2.3 The UV catastrophe. The solid line represents observed data which follows Planck's law. The dashed line represents the Rayleigh–Jeans law, which predicts a continuance to infinity.

A potential solution to the UV catastrophe arose from Max Planck's work, although he was not specifically seeking to solve the UV catastrophe as often suggested in textbooks.

Planck was interested in the relationship between energy and wavelength. He was able to intuitively modify Wien's law to a form which fitted the experimental curve everywhere and reduced to the Rayleigh–Jeans law at long wavelengths. Despite trying every method he could conceive of, Planck was unable to derive his equation from first principles which forced him to conclude that these first principles must be incorrect. According to classical physics, the electromagnetic radiation released from a black body source had an infinite number of modes of vibration[b] each with an average energy. Planck made a proposal that the opposite was true: the energy of each mode of vibration could only have a finite value that was proportional to frequency. To equate energy with frequency a constant of proportionality is required, which became known as the **Planck constant**, h:

$$E = h\nu = \frac{hc}{\lambda} \tag{2.1}$$

where $h = 6.63 \times 10^{32}\,\text{J}\cdot\text{s}^{-1}$. The implication of Eq. (2.1) is that energy is released (or absorbed) in multiples of the Planck constant, which each amount being referred to as a quantum of energy.

---

[b] Any complex body (i.e., anything other than a simple mass on a spring) can vibrate in a number of different ways, known as modes of vibration.

Understandably, Planck's quantum theory was met with some skepticism, but Albert Einstein was quick to accept the theory and apply it to his own work. At this time, Einstein was interested in an observation made some years earlier by Wilhelm Hallwachs (1859–1922) that when ultraviolet light struck a metal surface, rays were produced which behaved like cathode rays (which we now know are electrons). This phenomenon, which became known as the *photoelectric effect*, was further studied by Philipp Lenard (1862–1947) who demonstrated that the energy of the electrons produced was proportional to the frequency of the light. This appeared to be at odds with Maxwell's wave theory of light, which predicted that energy was proportional to radiance. Even more confusing was the observation that electrons only appeared to be ejected when the light reached a certain minimum frequency.

In *ca.* 1905, Einstein explained these observations by considering light not as a continuous wave (as classical theory states) but rather as a stream of "light packets," each with a certain amount of energy, consistent with Planck's theories on the quantization of energy. Einstein proposed that each quantum of light (called a *photon*) would have energy equal to its frequency multiplied by a constant, Eq. (2.1), and that only a photon above a certain minimum frequency would have enough energy to eject an electron from the metal. Einstein's work on the photoelectric effect was verified in 1915 when Millikan further characterized the properties of the electron, and Einstein was awarded the Noble Prize for Physics in 1921. The work of Planck and Einstein revolutionized physics and marked the emergence of quantum physics from which we have gained our current understanding of atomic structure.

2.6 The Bohr Model

Key Point: The energy of electrons is quantized which means they can only occupy specific regions of space known as orbits.

In the early decades of the 20th century, Rutherford's planetary model was the accepted atomic theory. However, this model presented a particular problem — according to classical physics, an orbiting electron is predicted to emit electromagnetic radiation as it moves, thus loosing energy. The implication of this is that the electron will have a decaying orbit, ultimately collapsing into the nucleus. This would mean that all atoms are unstable which was obviously incorrect. A second implication of this is that as the electron's orbit decays and it moves closer to the nucleus, the frequency of the electromagnetic radiation released should change continuously. This was at odds with experimental evidence which showed that atoms only emit light (electromagnetic radiation) at specific frequencies, giving rise to characteristic *line emission spectra*. The problems with the Rutherford atom were addressed by Niels Bohr in his landmark paper "On the Constitution of Atoms and Molecules" in 1923.

Bohr's first step was to evaluate the existing Rutherford model in terms of its strengths: the existence of the nucleus was irrefutable, as was the extra-nuclear position of the electrons. Bohr liked the idea of orbiting electrons and pictured the hydrogen atom as a tiny, one-planet solar system with the gravitational force of the solar system being replaced by the electrostatic force of attraction between oppositely charged particles. He borrowed Newton's law of universal gravitation and Coulomb's law and proposed that for an electron orbit to be stable, the centripetal force must be equal to the Coulombic force:

$$\frac{mv^2}{r} = \frac{Ze^2}{4\pi\varepsilon_0 r^2} \tag{2.2}$$

where Z is the atomic number and ε_0 is the permittivity of free space (Chapter 1). From this equation we see that

$$v^2 = \frac{Ze^2}{4\pi\varepsilon_0 mr} \tag{2.3}$$

According to classical physics, there are an unlimited number of possible values for v; Bohr, however, imposed a restriction known as the Bohr postulate which states that the angular momentum of the electron can only have values in multiples of $h/2\pi$, that is, $nh/2\pi$ where n is any integer. By introducing the Planck constant, Bohr was suggesting that the energy of the electron in the hydrogen atom was quantized and that it was held at a fixed position from the nucleus in a stationary state of constant energy. The angular momentum, L, of the electron is therefore quantized.

$$L = mvr = \frac{nh}{2\pi} \tag{2.4}$$

Combining Eqs. (2.3) and (2.4) and eliminating v gives an equation which will allow the radius of permitted orbital in the hydrogen atom to be calculated:

$$r = \frac{n^2 h^2 \varepsilon_0}{Ze^2 \pi m} \tag{2.5}$$

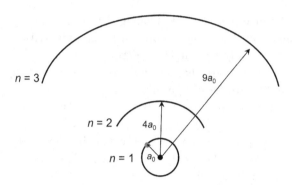

Figure 2.4 Bohr's stationary states in a hydrogen atom. Each orbital in a hydrogen atom is at a radius equal to $n^2 a_0$ ($a_0 = 0.0529 \, \text{nm}$).

Evaluation of the constants yields a more compact form in which a_o is known as the **Bohr radius**:

$$r = \frac{n^2}{Z} a_o \tag{2.6}$$

Bohr called each orbital a stationary state which is dependent on the value of n. For the hydrogen atom, $Z = 1$, and the spacing between each orbital is in multiples of $n^2 a_0$ (Figure 2.4).

Assuming that the angular momentum of the electron is quantized, it follows that the energy of the electron in a particular stationary state must also be quantized. The only mechanism for movement between these stationary states would be through the absorption or emission of a specific amount of energy equal to the difference in energy between the two states. Bohr felt this idea was consistent with the appearance of spectral lines and went on to describe the total energy of a stationary state as the sum of the kinetic and potential energies:

$$E = E_k + E_p = -\frac{Ze^2}{8\pi\varepsilon_0 r} \tag{2.7}$$

Using Eq. (2.5) to eliminate r, we arrive at an expression for the energy of a stationary state:

$$E = -\frac{me^4 Z^2}{8\varepsilon_0^2 n^2 h^2} \tag{2.8}$$

The difference in energy between two stationary states, ΔE, should be related to a frequency of electromagnetic radiation by Eqs. (2.1) and (2.8):

$$\Delta E = h\nu = E_{n_2} - E_{n_1} \tag{2.9}$$

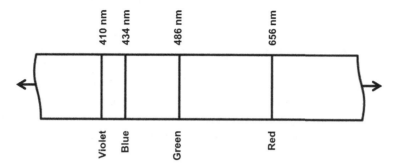

Figure 2.5 The Balmer series of the hydrogen emission spectrum. Further spectral lines are observed above and below the Balmer series.

which on expansion and rearrangement gives

$$\nu = \frac{me^4 Z^2}{8\varepsilon_0^2 h^3}\left(\frac{1}{n_1^2} - \frac{1}{n_2^2}\right) \quad \text{or} \quad \frac{1}{\lambda} = \frac{\nu}{c} = \frac{me^4 Z^2}{8\varepsilon_0^2 h^3 c}\left(\frac{1}{n_1^2} - \frac{1}{n_2^2}\right) \qquad (2.10)$$

The significance of this result is that it provides a link with Bohr's theoretical work with an experimentally observed phenomenon — line emission spectra for the hydrogen atom. These are observed when the light from a discharge tube filled with hydrogen is passed through a diffraction grating. A series of four discrete lines is observed in the visible region of the spectrum (Figure 2.5).

Prior to Bohr's work, several scientists including Johann Balmer (1825–1898) and Johannes Rydberg (1854–1919) sought to explain the pattern of spectral lines observed in hydrogen's emission spectrum. Eventually, Rydberg arrived at the empirical equation which bears his name:

$$\frac{1}{\lambda} = R_H\left(\frac{1}{n_1^2} - \frac{1}{n_2^2}\right) \qquad (2.11)$$

where R_H is the Rydberg constant $(1.097 \times 10^7 \, \text{m}^{-1})$ and n_1 and n_2 are integers with $n_1 < n_2$. Rydberg found that the wavelength of the four main spectral lines in the hydrogen emission spectrum could be predicted when $n_1 = 2$ and $n_2 = 3, 4, 5,$ or 6. Bohr's corollary of this is obtained when Eq. (2.10) is evaluated for $Z = 1$ and $n = 1$:

$$R_H = \frac{me^4}{8\varepsilon_0^2 h^3 c} = 1.097 \times 10^7 \, \text{m}^{-1}$$

We see that by taking appropriate values for the constants in Eq. (2.10), the Rydberg constant is returned and therefore Bohr's purely theoretical approach yields an identical result to that based upon experimental data.

One significant difference in the equations proposed by Bohr and Rydberg was that Bohr placed no restriction on the value of n. This suggested that there might be further spectral lines in the hydrogen emission spectrum that were yet to be observed experimentally. In 1916, Theodore Lyman (1874–1954) discovered a series of spectral lines in the far UV region and in 1922 Frederick Brackett (1896–1988) found a series in the IR region. A number of other series of spectral lines have since been discovered in other regions of the electromagnetic spectrum.

As noted earlier, Bohr's atomic model, although successful in some regards, was limited. It was found that it only applied to the hydrogen atom and could not explain the effect of an external magnetic field on spectral lines. Arnold Sommerfeld (1868–1951) attempted to modify Bohr's model by introducing the idea of elliptical orbits, but this approach was similarly unsatisfactory. Historically, we now regard Bohr's work as the "old quantum theory"; it was the subsequent work of Max Born (1882–1970), Werner Heisenberg (1901–1976) and Erwin Schrödinger (1887–1961) which led to the development of the atomic model we use today. Central to the success of this new atomic model was the idea that elementary particles such as electrons have wave-like properties just as electromagnetic radiation has particle-like properties.

2.7 Wave-Particle Duality

Key Point: All bodies have wave- and particle-like behaviors; as electrons exist at the quantum level, the wave nature is more pronounced leading to a degree of uncertainty in their properties.

In the early 1900 s, Einstein's interpretation of the photoelectric effect implied that light has some particle-like properties. Some years later, Arthur Compton (1892–1962) was to employ Einstein's ideas to explain an unexpected experimental finding. Compton was involved in work examining the scattering of X-rays. He found that when monochromatic X-rays were directed at a graphite target, the scattered beam had a slightly longer wavelength than the incident beam. The only explanation could be that some of the incident X-ray photons interacted with the material and lost a portion of their energy, increasing their wavelength. He also found that the loss in energy was dependent on the scattering angle. This phenomenon, known as the *Compton effect*, results from the inelastic scattering of a photon by an electron. If the incident photon is of lower energy, the photoelectric effect is more likely to occur, while at higher energies, the photon bombard the nucleus, causing formation of an electron and position through pair-production.

Compton's results, while significant in their own regard, were to have wider-reaching implications when Louis de Broglie (1892–1987) suggested that the opposite could also be true: particles can also have wave-like properties. To evaluate this, de Broglie derived an equation describing the wavelength of a particle based on Einstein's special theory of relativity:

$$E^2 = p^2c^2 + m^2c^4 \tag{2.12}$$

When the relativistic mass of the photon is zero, Eq. (2.12) reduces to Eq. (2.13) which is expanded to show the wavelength and frequency terms:

$$E = pc = p\nu\lambda \tag{2.13}$$

Taking into account Eq. (2.1) we see that the wavelength of a particle is given by the de Broglie relationship:

$$\lambda = \frac{h}{p} = \frac{h}{mv} \tag{2.14}$$

This relationship, proposed by de Broglie in 1924, was only partially supported by experimental evidence such as that for the particle-like nature of an electron, as shown by J.J. Thomson in 1904. It was in 1927 that Thomson's son, G.P. Thomson (1892–1975) together with Clinton Davisson (1881–1958) and Lester Germer (1896–1971) verified the wave-like nature of an electron. In their experiment, they observed that if electrons were accelerated by oppositely charged plates and directed toward a nickel crystal, a diffraction pattern was observed. They proposed that the diffraction pattern was as a result of electron waves being scattered by different layers of atoms in the metal crystal.

To get some idea of the magnitude of the wavelength of microscopic versus macroscopic particles, consider the de Broglie wavelength of an electron travelling at 1×10^7 m $\cdot$ s^{-1} and a 50 g tennis ball travelling at 90 miles per hour ($\sim$40 m $\cdot$ s^{-1}):

$$\lambda(\text{electron}) = \frac{6.6 \times 10^{-34}\,\text{kg} \cdot \text{m}^2 \cdot \text{s}^{-2}}{9.1 \times 10^{-31}\,\text{kg} \times 1 \times 10^7\,\text{m} \cdot \text{s}^{-1}} = 7.3 \times 10^{-11}\,\text{m}$$

$$\lambda(\text{tennis ball}) = \frac{6.6 \times 10^{-34}\,\text{kg} \cdot \text{m}^2 \cdot \text{s}^{-2}}{0.05\,\text{kg} \times 40\,\text{m} \cdot \text{s}^{-1}} \approx 1 \times 10^{-34}\,\text{m}$$

We see that the wavelength of the electron is roughly of the same order as the dimensions of an atom (1×10^{-12} m) which means that its wave-like properties will dominate its behavior at this level. Conversely, the wavelength of the tennis ball is insignificant when compared to its dimensions (*ca.* 7 cm), which means it's impossible to detect its wave behavior by any physical means.

The idea of wave-particle duality was not entirely consistent with the Bohr atom and de Broglie sought to reconcile the two. Taking the idea of Bohr's circular orbits and the wave-like nature of the electron, de Broglie realized that if the wavelength of the electron wave is multiplied by an integer, n, it will be equal to the circumference of a circle and constructive interference would occur. Destructive interference would occur if this condition is not satisfied, implying that n is a quantization condition. If this is true, then we would have

$$2\pi r = n\lambda \qquad (2.15)$$

$$2\pi r = \frac{nh}{mv} \quad \text{or} \quad mvr = \frac{nh}{2\pi} \qquad (2.16)$$

which is identical to Eq. (2.4) proposed by Bohr. Therefore, the de Broglie equation can be regarded as equivalent to a Bohr's equation and requires a quantization condition to be valid.

The idea that objects can have wave and particle-like properties is central to our understanding of quantum mechanics. In 1926, Max Born proposed that the amplitude of a wave at any point is related to the chance of finding the particle at that point; moreover, the square of the amplitude is directly proportional to the probability of finding the particle. This idea forms part of the *Copenhagen interpretation* of quantum mechanics which relates the properties of an electron to a mathematical expression known as a *wavefunction*. A further central tenet of quantum mechanics was proposed by Heisenberg in 1927. He suggested that it would be impossible to know both the position and momentum of a particle simultaneously. He was not able to prove this experimentally, but instead described a series of *gedankenexperimente*[c] which lead him to this conclusion.

Suppose there was an electron in orbit around a hydrogen nucleus. Any radiation which would be used to determine its position must be

---

[c]German "thought experiments."

of short wavelength, otherwise the position would not be well defined. However, short-wavelength photons will have high energy and will cause Compton scattering to occur, altering the electron's momentum. If we were to avoid the change in momentum by using long-wavelength photons, the position of the particle will be poorly defined. These conclusions are expressed as the **Heisenberg uncertainty principle** which can be written as

$$\delta x \delta p \approx \frac{h}{4\pi} \qquad (2.17)$$

where δx is the uncertainty in position and δp is the uncertainty in momentum. Since momentum is the product of mass and velocity, Eq. (2.17) may also be written as

$$\delta x \delta v \approx \frac{h}{4\pi m} \qquad (2.18)$$

Another form of Eq. (2.17) which is particularly relevant to nuclear chemistry is its expression in terms of energy and time. The uncertainty in time at which a particle exists at a particular position would be given by

$$\delta t = \frac{\delta x}{v} \qquad (2.19)$$

Since energy is $1/2mv^2$ and momentum is mv, the uncertainty in energy can be written as

$$\delta E \delta t \approx \frac{h}{4\pi} \qquad (2.20)$$

Equation (2.20) explains the uncertainty associated with radioactive decay. For example, if a nucleus is undergoing β-decay, the energy of the β-particles can be measured fairly precisely; however, this means that the time at which the emission occurs is very uncertain. This is often described as the stochastic nature of radioactive decay which means that decay processes are modeled using probability. Another significant consequence of the uncertainty principle is seen if we were to consider a particle confined to a very small space (small Δx). According to Eq. (2.20), we cannot know with certainty what the energy of the particle would be which implies that particle cannot have zero energy. The energy resulting from this restriction is known as the *zero-point energy*. This means that even at the absolute zero of temperature, an atom will retain some vibrational energy. This explains why liquid helium-4 will not freeze at atmospheric pressure regardless of how low the temperature drops (the zero-point energy is sufficient to maintain the liquid state).

2.8 The Schrödinger Model

Key Point: Due to the uncertainty associated with an electron's momentum and position, a probability-based approach employing wave mechanics can be used to describe atomic structure.

The implications of quantum mechanics for Bohr's atomic model were unavoidable, since his model was based on knowing the momentum and position of an electron this was in direct opposition to the uncertainty principle. In the mid-1920s, Heisenberg and Schrödinger independently discovered the underlying principles for a new kind of mechanics which incorporated the wave-particle duality of matter. Heisenberg referred to his approach as matrix mechanics, while Schrödinger's formalism was referred to as wave mechanics.

For chemists, Schrödinger's approach is synonymous with atomic structure and the mathematics involved is more familiar (though no less formidable). The central concept of Schrödinger's work was to provide a way of evaluating the wavefunction and total energy of a particle from knowledge of its kinetic and potential energies as a function of position. Although it is impossible to derive the Schrödinger equation from classical physics, we can borrow elements of it from the work of Einstein and de Broglie to arrive at a satisfactory derivation.

Starting with the equation for a travelling wave, Eq. (2.21), we differentiate (twice) with respect to x:

$$\psi(x,t) = a \sin\left[\frac{2\pi}{\lambda}(x - vt)\right] \tag{2.21}$$

$$\frac{d^2\psi}{dx^2} + \frac{4\pi^2}{\lambda^2}\psi = 0 \tag{2.22}$$

To incorporate wave-particle duality, we recall that the total energy of the particle is the sum of the kinetic and potential energies; therefore, by way of the de Broglie relation, we obtain an expression for λ^2:

$$E = \frac{p^2}{2m} + U \Leftrightarrow p = \sqrt{2m(E - U)} \tag{2.23}$$

$$\lambda^2 = \frac{h^2}{p^2} = \frac{h^2}{2m(E - U)} \tag{2.24}$$

Combining Eqs. (2.22) and (2.24), we obtain the ***time-independent Schrödinger equation*** in one-dimension:

$$\frac{\mathrm{d}^2\psi}{\mathrm{d}x^2} + \frac{8\pi^2 m(E-U)}{h^2}\psi = 0 \tag{2.25}$$

Employing the relation for the reduced Planck constant, $\hbar$, and rearranging to collect together the kinetic and potential energy terms, we obtain a simplified expression:

$$\text{Let } \hbar = \frac{h}{2\pi} \quad \therefore \quad -\frac{\hbar^2}{2m}\frac{\mathrm{d}^2\psi}{\mathrm{d}x^2} + U\psi = E\psi \tag{2.26}$$

Equation (2.26) can be presented in a much more compact way through the use of the ***Hamiltonian operator***, $\hat{H}$, which is the sum of the kinetic and potential energies:

$$\hat{H} = -\frac{\hbar^2}{2m}\frac{\mathrm{d}^2\psi}{\mathrm{d}x^2} + U \tag{2.27}$$

For a particle in three dimensions, we must include y and z terms in Eq. (2.26), which modifies the Hamiltonian by way of the Laplacian operator, ∇^2 ("del-squared"):

$$\text{Let } \nabla^2 = \frac{\partial^2}{\partial x^2} + \frac{\partial^2}{\partial y^2} + \frac{\partial^2}{\partial z^2} \quad \therefore \quad \hat{H} = -\frac{\hbar^2}{2m}\nabla^2\psi + U\psi \tag{2.28}$$

Overall, this gives us a very compact form of the time-independent Schrödinger equation in three dimensions:

$$\hat{H}\psi = E\psi \tag{2.29}$$

Fundamentally, the Schrödinger equation is a wave equation and is subject to the same basic principles outlined in Chapter 1. The travelling wave described by Eq. (2.21) can continue to propagate through space, as could the matter wave described by Schrödinger. However, recall that when we imposed boundary conditions on Eq. (2.21) (presented in Chapter 1, Figure 1.13 *et seq.*), we found that a standing wave was obtained. A similar idea is applied to the Schrödinger equation and discrete values of E are only obtained when the electron is constrained to move in a defined region of space described by boundary conditions. These discrete energy values are known as eigenvalues and the corresponding wavefunctions are known as ***eigenfunctions***; only an allowed wavefunction, when acted on by the

Hamiltonian operator, returns the same wavefunction multiplied by E. This implies that certain wavefunctions are associated with specific energy levels.

If we consider a one-electron system (H, He$^+$, Li^{2+} etc.) the Schrödinger equation would be

$$-\frac{h^2}{8\pi^2\mu}\nabla^2\psi - \frac{Ze^2}{4\pi\varepsilon_0 r}\psi = E\psi \tag{2.30}$$

where μ is the reduced mass of the electron (this takes into account the pull of the nucleus). The potential energy term arises entirely from the Coulombic attraction of the nucleus, which creates a spherically symmetrical potential field; because of this, it is more convenient to use spherical polar coordinates rather than Cartesian coordinates; in this case x, y, and z are replaced by r, θ, and ϕ. To solve the Schrödinger equation for the hydrogen atom, Eq. (2.30) is separated into three equations so that the wavefunction is represented by the product of these three equations:

$$\psi(r, \theta, \phi) = R(r)P(\theta)F(\phi) \tag{2.31}$$

(1) The radial equation, $R(r)$, requires a constant known as the **principle quantum number**, n, which governs the energy of the electron. The allowed values are $n = 1, 2, 3 \ldots \infty$ such that as the radius of the orbital increases, so too does the value of n.

(2) The colatitude equation, $P(\theta)$, requires a constant, the **angular momentum quantum number**, l, and this describes the shape of the atomic orbital. It has allowed values $l = 0, 1, 2 \ldots (n-1)$.

(3) Finally, solution of the azimuthal, $F(\phi)$, equation requires the third constant, the **magnetic quantum number**, m_l, which is constrained to take on $2l + 1$ values ($m_l = 0, \pm 1, \pm 2 \ldots \pm l$).

The solutions to the colatitude and azimuthal equations are often considered together as the **angular part** of the wavefunction, $Y(\theta, \phi)$, while the solution of the radial equation is considered to be the **radial part** of the wavefunction.

For a hydrogen-like atom, the lowest energy state has $n = 1$ which gives $l = 0$ and $m_l = 0$, and we see that the wavefunction is only dependent on the radial equation; it is independent of latitude and longitude. This is referred to as a 1 s orbital and by employing the Born interpretation of the wavefunction, we could view this graphically as having a spherical boundary surface. When $n = 2$ and $l = 0$ we also have a spherical orbital, this time with a boundary surface farther from the nucleus. However, when

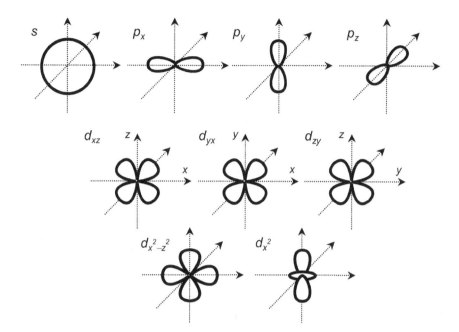

Figure 2.6 Shapes of atomic orbitals. The three-dimensional shapes represented the boundary surfaces predicted by the Schrödinger equation.

$n = 2$, l can also be 1, which leads to three possible values for m_l (-1, 0, and $+1$). This is known as a 2p orbital and it is divided up into three sub-orbitals, $2p_x$, $2p_y$, and $2p_z$, depending on the value of m_l. A summary of the quantum numbers and the orbitals they give rise to is given in Figure 2.6 and Table 2.1.

While this description of the atom appeared to take care of some inconsistencies which arose from the Bohr model, it could not explain certain other phenomena, such as the *fine structure* of the hydrogen emission spectrum. At first glance, the emission spectrum for hydrogen has a single line at *ca.* 656 nm, which is consistent with the Schrödinger (and Bohr) model. However, when the single line at 656 nm is viewed at higher resolution it appears as a closely spaced doublet which cannot be explained by the Schrödinger equation alone. If the emission spectrum of deuterium is similarly examined, an identical doublet is observed. Since it is only the electronic structure which is shared between hydrogen and deuterium, this similarity in the fine structure must arise from a property of the electron. As hydrogen and deuterium have an identical set of quantum

Table 2.1 Summary of the Quantum Numbers (up to $Z = 56$)

Shell	n	l	m	Subshell	No. of Electrons
K	1	0	0	1s	2
L	2	0	0	2s	2
	2	1	$-1, 0, +1$	2p	4
M	3	0	0	3s	2
	3	1	$-1, 0, +1$	3p	4
	3	2	$-2, -1, 0, +1, +2$	3d	10
N	4	0	0	4s	2
	4	1	$-1, 0, +1$	4p	6
	4	2	$-2, -1, 0, +1, +2$	4d	10
	4	3	$-3, -2, -1, 0, +1, +2, +3$	4g	14

numbers, this in turn implies that a fourth quantum number is required to explain the difference.

Wolfgang Pauli (1900–1958) tackled the significance of this fourth quantum number by proposing that two identical electrons cannot exist in the same quantum state simultaneously. Applying this to electronic structure, this means than if two electrons are in the same atomic orbital, the values of n, l, and m_l will be the same, but the fourth quantum number must be different. This is the **Pauli exclusion principle**. The clue to the exact nature of this fourth quantum number came from the work of Otto Stern (1888–1969) and Walther Gerlach (1889–1979) in 1922. In their experiment, a stream of silver atoms was directed through a heterogeneous magnetic field. As silver atoms have one valence electron with $l = 0$ (s-orbital), there should be no interaction with the magnetic field. This would mean that the atoms should form a continuous band on the detector screen. However, what was actually observed was the deflection of the atoms into two distinct groups. This implied that there was some property of the valence electron which imparted a magnetic moment.

In 1925, George Uhlenbeck (1900–1988) and Samuel Goudsmit (1902–1978) proposed that the existence of two groups of electrons could be explained by assigning an intrinsic angular momentum to the electron, referred to as **spin**, s. In essence, they said that as moving charges exert a magnetic moment, and electrons have been shown to possess both charge and a magnetic moment, they must have intrinsic angular momentum (spin). Since solving the Schrödinger equation produces a restriction on angular momentum of $(2l + 1)$, analogously the intrinsic angular momentum of an electron was conceived to have $(2s + 1)$ orientations. This means that given the result of the Stern–Gerlach experiment (two groups of orientation) we have $(2s + 1) = 2$ which gives $s = 1/2$. To distinguish between the two possible orientations, an electron can have $s = +1/2$ (spin "up") or $s = -1/2$ (spin "down"), as shown in what have become known as Pauli diagrams.

2.9 The Relationship between the Nucleus and Electrons

Key Point: The presence of charged moving particles in the nucleus produces a magnetic field which interacts with orbiting electrons.

In our discussion of atomic structure so far we have largely neglected the role of the nucleus in the electronic structure of the atom. We have implicitly assumed that the nucleus has infinite mass, that it remains stationary and occupies a point in space. Of course, these assumptions are incorrect. The nucleus has a finite mass governed by its composition and will therefore move along with the orbiting electrons according to their common center of gravity. The nucleus also contains subatomic particles which possess spin and therefore have a magnetic moment. Consequently, we would expect the nucleus to have an impact on the energy of orbiting electrons.

One of the most significant phenomena resulting from an interaction between the nucleus and electrons of an atom is the *hyperfine structure*. This is a splitting of electronic energy levels due to the interaction of the nuclear magnetic moment with the magnetic field produced by orbiting electrons. The nuclear magnetic moment arises from the intrinsic magnetic moment of the protons and neutrons (both $1/2$-spin particles). As spinning fermions tend to pair up with another similar particle of opposite spin, nuclei with an even number of protons and an even number of neutrons will have zero overall magnetic moment (so-called even–even nuclei). The odd–odd or odd–even nuclei can have integer or half-integer spin and these nuclei will have an overall magnetic moment.

An interesting point to note at this stage is that the Rydberg constant must vary from element to element, as the mass of the electron is affected by the nucleus. In this case, we employ the reduced mass which is the product over the sum of the two masses. This enabled Harold Urey (1893–1981) to first identify deuterium spectroscopically by taking into account the redshift predicted for the Balmer series. Urey was awarded the Noble Prize for Chemistry in 1934 for his discovery of "heavy hydrogen."

Chapter Summary

- Our understanding of the structure of the atom has come from experimental evidence collected over two centuries. Although now disproved, much of the early work conducted by Rutherford, Thomson *et al.* provided the first evidence for subatomic structure.
- A major milestone in the development of modern atomic theory was the proposition of the quantum by Max Planck. This not only resolved the UV catastrophe but importantly predicted features of atomic structure that were yet to be discovered.
- The de Broglie–Bohr atom represented the first reconciliation of Planck's quantum theory with models of the atom. This model could satisfactorily explain the existence of spectral lines in the atomic emission spectrum of hydrogen.
- The introduction of the Heisenberg uncertainty principle marked the start of the new quantum theory, which was dominated by Heisenberg and Schrödinger, the latter using wave mechanics to describe the position of electrons using probability.
- The existence of spin was incorporated into the quantum theory by Pauli. This would have significant implications for model of atomic and nuclear structure, and explained the existence of hyperfine interactions.

Review Questions

(1) Consider Thomson's cathode ray experiment. If the tube was evacuated and then filled with hydrogen, what would be the likely observations?

(2) In the photoelectric effect, incident photons must possess minimum threshold energy. If silver's threshold energy is $4.73\,\text{eV}$, what is the minimum wavelength which can eject an electron? Take $h = 6.62 \times 10^{-34}\,\text{J}\cdot\text{s}$, $c = 2.99 \times 10^8\,\text{m}\cdot\text{s}^{-1}$ and $1\,\text{eV} = 1.60 \times 10^{-19}\,\text{J}$.

(3) What wavelength of light would be observed in an electronic transition $n_3 \rightarrow n_2$. Take $R_\text{H} = 1.097 \times 10^7\,\text{m}^{-1}$.

(4) Suggest how knowledge of the Rydberg equation could help astronomers detect specific elements in the Earth's Sun.

(5) Calculate the de Broglie wavelength of a $70\,\text{kg}$ man walking at an average speed of 3.1 miles per hour. Take $h = 6.62 \times 10^{-34}\,\text{J}\cdot\text{s}$ and one mile per hour as equivalent to $0.447\,\text{m}\cdot\text{s}^{-1}$.

(6) If a radionuclide undergoes beta decay at a known instant in time, how might you expect the peak on an energy spectrum to appear?

(7) Qualitatively compare the uncertainty in position, δx, for an electron travelling at 1×10^7 m·s^{-1} with that of the 70 kg man in the previous question.

(8) The radial part of the wavefunction for a 1s orbital can be written as $\psi \propto e^{-r/a_0}$. Based on this relationship, how would you expect the wavefunction to vary with radius, r?

(9) What shape would the valance orbital be in a potassium ion, K$^+$? State the electronic configuration of the ion.

(10) For the $n = 5$ state, what are the allowed values of the angular momentum quantum number and the magnetic quantum number?

Chapter 3

The Structure of the Nucleus

"So, naturalists observe, a flea
Has smaller fleas that on him prey;
And these have smaller still to bite 'em,
And so proceed *ad infinitum*."

J. Swift

Since the discovery of radioactivity in 1896 by Henri Becquerel, the atomic nucleus has captivated scientists from many disciplines. Their pioneering work and that of many others lead to the splitting of the atom in the early 20th century and has brought us closer to reducing our dependence on fossil fuels. On completion of this chapter and the associated questions, you should:

- Be familiar with the terminology associated with nuclear structure.
- Be able to calculate the binding energy for a particular nuclide.
- Be able to approximate the volume and density of the nucleus of an atom using the liquid-drop model.
- Be able to interpret the shell model using the idea of magic numbers.

3.1 Nuclides

Key Point: The combination of protons and neutrons in different proportions in a nucleus provides a means of classifying nuclides.

The combination of protons and neutrons in the nucleus of an atom gives rise to a particular nomenclature for *nuclides*. A nuclide is an atomic species which is characterized by the contents of its nucleus (protons and neutrons) and its energy state (ground state or excited state). When a set of nuclides have the same number of protons but a different number of neutrons, they are referred to as *isotopes* (e.g., $^{12}_{6}C$ and $^{13}_{6}C$). Conversely, when a set of nuclides have the same number of neutrons but a different number of protons they are known as *isotones* (e.g., $^{13}_{6}C$ and $^{14}_{7}N$). When a set of nuclides have the same mass number but different atomic number, they care called *isobars* (e.g., $^{17}_{7}N$ and $^{17}_{8}O$).

We describe a stable nuclide as one which does not undergo radioactive decay. There are currently believed to be 254 stable nuclides, representing 80 of the known elements. The majority of these are *primordial nuclides*; that is, they have existed since the formation of the solar system (in fact, there are a further 34 nuclides which are classified as primordial as they have $t_{1/2} > 5 \times 10^7$ years).

3.2 The Nucleons

Key Point: Protons and neutrons, collectively called the nucleons, are subatomic particles, which are composed of fundamental particles known as quarks.

In Chapter 2, we introduced the Geiger–Marsden (gold-leaf) experiment in which helium nuclei (α-particles) were directed toward thin sheets of gold. In most observations, the α-particles passed straight through the gold leaf; however, some were deflected at angles of $180°$, suggesting a head-on collision with a similarly charged body. It was Ernst Rutherford who successfully interpreted this data by adopting a model in which the α-particles were deflected by the Coulomb repulsion of a central region in the atom. He deduced that the kinetic energy of the incident α-particle must be completely converted to potential energy at the point of closest approach; given that the α-particles had energy of 7 MeV, he was able to calculate this distance using Coulomb's law (*c.f.* Chapter 1):

$$\frac{1}{2}mv^2 = k\frac{q_1 q_2}{r} = k\frac{(2e)(Ze)}{d} \tag{3.1}$$

$$d = \frac{4kZe^2}{mv^2} = 3.2 \times 10^{-14} \text{ m} \tag{3.2}$$

When Rutherford repeated the experiment with other metals (e.g., silver), he obtained results with a similar magnitude and concluded that the positive charge must be concentrated in a small sphere with a radius *ca.* 10^{-14} m, which he named the nucleus.

The next challenge for Rutherford was to discover what made the nucleus of one atom different from another. In 1917, he demonstrated that it was possible to convert at atom of nitrogen into an atom of oxygen by bombarding it with helium nuclei:

$$^{14}_{7}\text{N} + ^{4}_{2}\text{He} \rightarrow ^{17}_{6}\text{O} + ^{1}_{1}\text{H}$$

The release of hydrogen suggested that the nuclei of all atoms contained a hydrogen nucleus. Further experiments verified this to be the case and in 1920, Rutherford proposed the name ***proton*** to distinguish the hydrogen nucleus from nascent hydrogen.

In the contemporary context, protons are classified as hadrons according to the Standard Model and are composed of three quarks held together by the strong force which is mediated by gluons (Figure 3.1). The properties of the proton (Table 3.1) are, therefore, governed by its constituents. Modern measurements have shown that only 1% of the measured mass of a proton arises from the rest mass of the constituent quarks; the remaining 99% arises from the kinetic energy of the constituent particles (*c.f.* Einstein's special relativity).

Based on the discovery of the proton, it was known that a nucleus with Z protons would have a charge of $+Ze$ and a mass of *ca.* $2Z$. To account for the mass in the nucleus, Rutherford proposed the idea of the ***neutron*** as an electrically neutral counterpart to the proton. However, it was James Chadwick (1891–1974), working with Rutherford in the 1930s, who provided the first experimental proof of the neutron. In his experiments, Chadwick directed alpha particles from a sample of polonium toward a beryllium

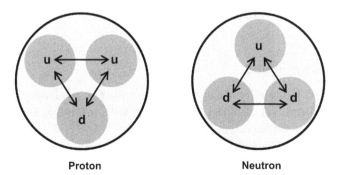

Proton　　　　　　　　　　**Neutron**

Figure 3.1 Structure of a proton and neutron. According to the standard model, proton and neutrons are classified as hadrons and contain up quarks and down quarks held together by gluons.

Table 3.1 Properties of Protons and Neutrons

	Proton	Neutron
Symbol	p^+ or $^1_1 p$	n^0 or $^1_0 n$
Classification	Hadron	Hadron
Nest mass (kg)	1.6726×10^{-27}	1.6749×10^{-27}
Charge (C)	1.6022×10^{-19}	0

target. When the alpha particles struck the beryllium,[a] a stream of particles was produced which were directed toward a paraffin block. Paraffin, being hydrogen-rich, when struck by these particles, released protons which were detected by an ionization chamber. The range and mass of the particle appeared almost identical to that of the proton, but it was electrically neutral. Chadwick named this particle the neutron and he was awarded the Noble Prize for Physics in 1935 for his discovery.

Neutrons, like protons, are composed of quarks held together by gluons. However, they have a slightly larger mass than protons and have a zero electrical charge. Unlike protons, free neutrons are unstable; they undergo radioactive decay within *ca.* 15 min becoming a proton and releasing an electron (β-particle) and an electron neutrino:

$$^1_0 n \rightarrow {}^1_1 p + e^- + \bar{\nu}_e$$

A fraction of free neutrons (*ca.* 1/1000) will also produce a gamma ray photon on decay:

$$^1_0 n \rightarrow {}^1_1 p + e^- + \bar{\nu}_e + \gamma$$

Neutrons bound within the nucleus are generally stable; in unstable nuclei (those which undergo radioactive decay), the neutron decays are as shown above.

3.3 Nuclear Stability

Key Point: The stability of a nucleus is controlled by a balance of opposing forces. The formation of nuclei is associated with a binding energy which obeys the law of conservation of energy.

Not long after Chadwick's discovery of the neutron, Werner Heisenberg proposed the **proton–neutron model** of the nucleus, in which the

---

[a]These particles (neutrons) bring about the conversion of beryllium to carbon according to the equation $^4_2 He + {}^9_4 Be \rightarrow {}^{12}_6 C + {}^1_0 n$.

protons and neutrons experienced a short-range binding force in addition to the Coulomb repulsive force between protons. We can estimate this latter property for a simple two-proton system using a modification of Coulomb's law:

$$U = \frac{e^2}{4\pi\varepsilon_0 r} \left[\frac{Z(Z-1)}{2} \right] \tag{3.3}$$

The term in square brackets gives the number of proton pairs for Z protons. For helium, $Z = 2$ and $r = 2 \times 10^{-15}$ m. Therefore, we obtain:

$$U = \frac{(1.602 \times 10^{-19})^2}{4\pi \times (8.854 \times 10^{-12}) \times (2 \times 10^{-15})} \cdot \frac{2(2-1)}{2} = 1.153 \times 10^{-13} \text{ J}$$

which is equivalent to 0.72 MeV. If we compare this to the value obtained for a heavy element, say uranium ($Z = 92$, $r = 8 \times 10^{-15}$ m), the result is 750 MeV. This implies that as Z increases, nuclei would be expected to become more unstable. This in fact is true and we will see in later chapters that there are no stable nuclei having $Z > 83$.

When nucleons combine there is an associated ***nuclear binding energy***. Consider the formation of deuterium, an isotope of hydrogen with one proton and one neutron. At first glance, you could be forgiven for thinking that the mass of a deuteron should be the sum of the individual masses of a proton and a neutron, that is,

$$m_d = m_p + m_n = 1.007277 + 1.008665 = 2.015942 \text{ u} \tag{3.4}$$

It seems, though, that when the actual mass of a deuteron is measured, a value lower than that predicted is obtained ($m_d = 2.013553$ u). This discrepancy of some 0.002389 u is accounted for by the release of a gamma ray photon during the formation of deuterium:

$$^1_1\text{H} + ^1_0\text{n} \rightarrow ^2_1\text{H} + \gamma$$

The energy of the gamma ray photon is 2.225 MeV, which is equivalent to a mass of 0.002389 u.[b] Thus, we see that the formation of deuterium obeys the law of conservation of energy. We define the nuclear binding energy as

---

[b]One atomic mass unit (u) is equivalent to 931.48 MeV or 1.602×10^{-13} J. Thus, the rest mass of a hydrogen atom, $m_H c^2 = 1.007825 \times 931.48 = 938.77$ MeV.

the reverse of this process — the energy required to break apart a nucleus:

$$BE = (mc^2)_{\text{pieces}} - (mc^2)_{\text{nucleus}} \qquad (3.5)$$

It follows that the greater the repulsive force between nucleons, the lower the binding energy. The nuclear binding energy can be conveniently calculated using the atomic number and atomic mass of the atom, and the mass of a hydrogen atom:

$$BE = Zm_Hc^2 + Nm_nc^2 - m_xc^2 \qquad (3.6)$$

It is also useful to consider the **binding energy per nucleon**, which when plotted against the number of nucleons (Figure 3.2) shows that for stable nuclei with $A > 20$, the binding energy per nucleon is fairly constant at around 6 MeV. This is in keeping with the idea that the forces between nucleons are **saturated** — that is, a specific nucleon only forms attractive "bonds" with a limited number of other nucleons. As the nuclear force is only effective over a short range, the nucleons must be close together, suggesting a close-packed model of the nucleus.

While it is obvious that the nuclear force must be greater than the Coulomb repulsive force, it is less obvious why the nuclear force is only effective over a short range. To explain this latter force, we must rely on data from neutron scattering experiments. If neutrons are directed at a target containing hydrogen, the resultant elastic neutron scattering can be used to determine the potential energy of the system. When this is set

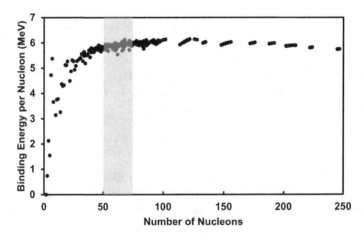

Figure 3.2 Binding energy per nucleon versus number of nucleons. The region of greatest stability is around 60; nucleons with $A > 60$ are not as tightly bound and, therefore, less stable.

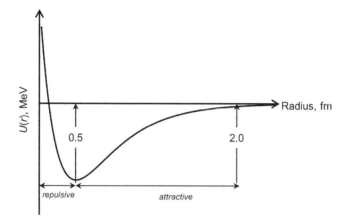

Figure 3.3 Potential energy versus nucleon separation. The depth of the well indicates a repulsive force, which prevents nucleons from approaching much closer than 0.4 fm.

against distance (Figure 3.3) we see that up to a distance of about 0.5 fm the nuclear force is repulsive; this keeps nucleons separated and is evidence for the Pauli exclusion principle in nuclei. Beyond this up to about 2 fm the force is attractive, holding nucleons together; it then decays asymptotically to negligibly small values.

An important question is what "is the origin of the nuclear force?" We mentioned in Chapter 1 that the nuclear force is believed to be a residue of the strong interaction within nucleons. It was Hideki Yukawa (1907–1981) who first described the interaction between nucleons using an **exchange force model**. According to this model, two particles — such as a proton and a neutron — experience a force of attraction when one particle spontaneously releases a particle and the other absorbs it. We have already seen in Chapter 1 that such process can be represented using a Feynman diagram (Figure 3.4). At first glance, this idea would seem to violate the conservation of energy — for example, a proton of defined mass releases energy (mc^2) yet remains a proton, despite having seemed to have lost mass. The explanation comes from the idea of a **virtual particle**. A virtual particle is one which cannot be directly detected during an energy-violating process, but can be inferred from the Heisenberg uncertainty principle. For example, the uncertainty principle for a virtual particle could be expressed as:

$$\delta E \cdot \delta t \geq \frac{\hbar}{2} \tag{3.7}$$

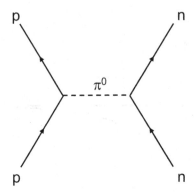

Figure 3.4 Feynman diagram for the exchange force model. A pi meson (pion) mediates the nuclear force.

Equation (3.7) implies that energy conservation can be violated by an amount δE for a time δt. Since the amount of energy exchanged in Figure 3.4 is given by mc^2 we can state

$$\delta t \approx \frac{\hbar}{\delta E} = \frac{\hbar}{mc^2} \tag{3.8}$$

Given that the maximum distance, d, between the proton and neutron must be related to the time interval by the speed of light, $d = c\delta t$, we can approximate the rest energy of the exchange particle:

$$\delta E = \frac{\hbar c}{d} = \frac{(1.1 \times 10^{-34}) \cdot (3.0 \times 10^8)}{(2.0 \times 10^{-15})} = 1.6 \times 10^{-11}\,\mathrm{J} = 100\,\mathrm{MeV}$$

In the late 1940s, the **pi meson** was discovered which appeared to fit the criteria of the exchange particle for the nuclear force.

3.4 The Liquid-Drop Model

Key Point: The binding energy of a nucleus can be predicted by a semi-empirical equation which assumes the nucleus has a shape like that of a drop of liquid.

In 1935, the German physicist Carl von Weizsäcker (1912–2007) published the first complete model of the atomic nucleus. This model was based on an earlier idea proposed by Niels Bohr, and described the protons and neutrons in a nucleus as an incompressible fluid. While this model can be used to explain many nuclear properties, it ultimately fell short of complete agreement with experimental data. However, this model was sufficiently accurate to enable the start of the nuclear arms race in the 1940s and the events leading up to the detonation of the first thermonuclear weapon on July 16, 1945 at the Trinity test site in New Mexico.

Key to the liquid-drop model was being able to define the dimensions of a nucleus. The gold-leaf experiment showed that the volume occupied by the nucleus of an atom was incredibly small — about 10^{-15} m. Using this scale and assuming a close-packing model of the nucleus, we can expect the size of the nucleus to be directly proportional to the number of nucleons present, represented by A, the mass number. If we assume that the nucleus is spherical, and that the volume of a sphere is proportional to its radius cubed, we can conclude that the radius of a nucleus would be given by

$$r = r_0 A^{1/3} \tag{3.9}$$

The proportionality constant, r_0, can be determined from a plot of $\ln A$ versus $\ln r$ for a range of nuclei. For the data shown in Figure 3.5, the linear regression equation is $\ln A = 0.3331 \cdot \ln r - 34.489$ and the y-intercept is, therefore, $e^{-34.489} = 1.05 \times 10^{-15}$ m which is in good agreement with the accepted value $(1.4 \times 10^{-15}$ m$)$.

If we consider the nucleus as a closely packed sphere, we can calculate the nuclear density as its mass per unit volume. So, for example, the density of a gold nucleus is 1.46×10^{13} kg $\cdot$ m^{-3} and that of a carbon nucleus is 1.47×10^{13} kg $\cdot$ m^{-3}. The fact that these two values are almost identical suggests that nuclear density is independent of mass number. We can show this by first expressing the volume of a sphere in terms of the proportionality constant, r_0:

$$v = \frac{4}{3}\pi r^3 = \frac{4}{3}\pi (r_0 A^{\frac{1}{3}})^3 \tag{3.10}$$

Since the mass of a nucleus is the product of the mass number and 1/12th the mass of carbon-12 we get:

$$m = uA \quad \therefore \ \rho = \frac{uA}{\frac{4}{3}\pi r_0^3 A} = \frac{u}{\frac{4}{3}\pi r_0^3} \tag{3.11}$$

This result implies that most of an atom's mass is found in its nucleus, which is consistent with Rutherford's findings in the gold leaf experiment.

When nuclei were investigated by bombardment with high-energy electrons, it was discovered that they do not possess uniform density; instead, the density decreases slightly toward the outer edge of the nucleus. This picture of the nucleus brings to mind the image of a drop of liquid.

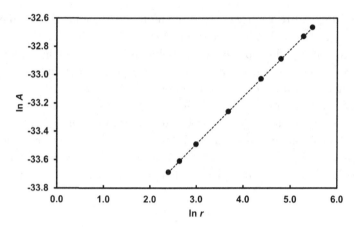

Figure 3.5 Determining r_0. The proportionality constant in Eq. (3.9) shows the dependence of radius on nucleon number.

In such a model, the binding energy of the nucleus will depend on five factors:

(1) Volume: the binding energy is proportional to the number of nucleons, which is in turn proportional to the volume of the nucleus.
(2) Surface: nucleons closer to the surface of the nucleus will have fewer neighbors; therefore, these will reduce the binding energy.
(3) Coulomb repulsion: each proton repels every other proton by an amount proportional to the Coulomb potential, reducing the binding energy.
(4) Asymmetry: the Pauli exclusion principle restricts how many fermions can occupy a quantum state. As more are added, they must occupy higher energy levels, decreasing the binding energy.
(5) Neutron excess: for heavy nuclei, the binding energy is reduced by an amount related to the number of excess neutrons.

These four factors form the basis of the ***semi-empirical binding energy*** formula, proposed by Weizsäcker in 1935:

$$\mathrm{BE} = c_1 A - c_2 A^{\frac{2}{3}} - c_3 \frac{Z(Z-1)}{A^{\frac{1}{3}}} - c_4 \frac{(N-Z)^2}{A} - \delta(A, Z) \qquad (3.12)$$

For nuclei with $A \geq 15$, the first four empirical constants take values of 15.7, 17.8, 0.71, and 23.6 MeV, respectively. The final term, $\delta(A, Z)$, varies: for even–even nuclei, +11.18; for odd–odd, −11.18; and for odd–even it is zero.

The predicted binding energies (using the Weizsäcker formula) are in good agreement with those determined experimentally, which supports the idea of a liquid-drop model. However, this model could not explain certain experimental evidence, most notably the nuclear magnetic moment, and it became clear that another model would be required.

3.5 The Shell Model

Key Point: The quantization of nucleons permits a description of nuclear structure using a shell model. This model can be used to explain the stability of certain nuclides using the so-called magic numbers.

The Bohr and Schrödinger models of the atom were so successful in explaining various chemical phenomena that they became an obvious template for constructing a model of the nucleus. So, for example, just as the Bohr and Schrödinger models predict that a closed-shell configuration confers electronic stability (*viz.* the noble gases with 2, 10, 18, 36, 54, and 86 electrons), we would expect stable nuclei to have an analogous closed-shell configuration. This idea was first suggested by James Bartlett (1904–2000) who observed that even–even nuclei have increased stability, and those nuclei with N or Z equal to 2, 8, 20, 28, 50, 82, and 126, the so-called ***magic numbers***, are particularly stable.[c] Bartlett believed this evidence was sufficiently compelling to support the concept of a shell model of the nucleus.

The shell model is based on the idea that each nucleon moves around independently of other nucleons in a well-defined ***potential well*** with potential energies as a function of the distance from the center of the nucleus. This means that the energy of the individual nucleons is quantized, and that the energy levels are solutions of the Schrödinger equation. We see that the potential energy term of a nucleon in the Schrödinger equation varies with radius according to the Woods–Saxon potential, given by

$$V(r) = \frac{V_0}{1 + e^{\left(\frac{r-R}{a}\right)}} \tag{3.13}$$

where V_0 is the potential well depth (*ca.* 50 MeV), r is the distance from the center of the nucleus, R is the nuclear radius from Eq. (3.9) and a is a constant (*ca.* 0.5 fm). If we were to interpret the Woods–Saxon potential

---

[c]Some elements, e.g.,^{40}ca., are said to be doubly magic, as both N and Z are equal to a magic number.

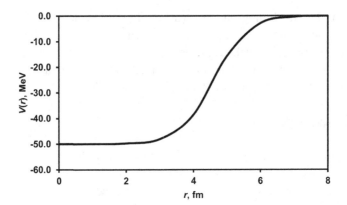

Figure 3.6 Woods–Saxon potential. As the nucleon gets closer to the surface of the nucleus, its potential energy decreases.

graphically as a function of radius (Figure 3.6), we would see that as the nucleon gets closer to the surface of the nucleus, the potential falls to almost zero (i.e., the nucleon is lower in energy).

The quantization of nucleons means that each nucleon is described by a set of four quantum numbers, as was the case for an electron. In fact, the quantum numbers used to describe nucleons are not particularly different from those used to describe electrons, except that the angular momentum quantum number, l, is not constrained to $n-1$ values; rather it can take all positive integers starting with zero.[d] As with electrons, no two nucleons can have exactly the same set of quantum numbers, so the Pauli exclusion principle must also apply to nucleons. In 1949, Maria Goeppert-Mayer proposed that in nuclei the angular momentum quantum number would combine with the spin angular momentum quantum number, s, to give the **total angular momentum quantum number**, j

$$j = \ell \pm s \tag{3.14}$$

such that there would be $2j + 1$ possibilities, referred to as **multiplicities**. In other words, the value of j describes how many nucleons can occupy a particular state ("orbital"). In the simplest example (the first state), we

---

[d]The principle quantum number, n, does not have the same meaning for nucleons in the shell model as it does for electrons in a hydrogen-like atom. It is still required to describe the wavefunction for the nucleon, but is more like a radial quantum number.

would have

$$\ell = 0, \; j = 1/2$$

giving rise to a state which two nucleons can occupy $(2 \times 1/2 + 1)$. To identify the shells, we use the same labeling system as for electronic orbitals ($l = 0$ is s, $l = 1$ is p, $l = 2$ is d, $l = 3$ is f, etc.) So, the state in the example above would be identified as $1s_{1/2}$. In the next case, we could have

$$\ell = 1, j = 3/2, \quad \text{and} \quad 1/2$$

which gives a total multiplicity of six (four for $j = 3/2$ and two for $j = 1/2$). The two states would be identified as $1p_{3/2}$ and $1p_{1/2}$. If we continue this process and draw a type of energy level diagram we see that a pattern emerges, in which the magic numbers appear (Figure 3.7). This is analogous to how the Aufbau method for filling electronic energy levels shows that the noble gases have a full electronic state, which is associated with their stability.

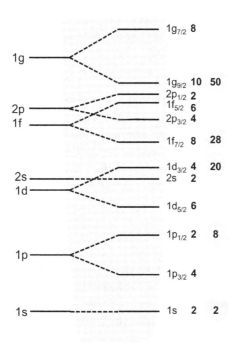

Figure 3.7 Nuclear shell model. Nucleons are placed in quantized energy levels within a nuclear potential well with walls given by the Woods–Saxon potential. Numbers in grey circles are the so-called magic numbers.

The shell model can be used to explain an important property of nuclei, that of **nuclear spin**, I, which is the basis of nuclear magnetic resonance (NMR) spectroscopy. Nuclear spin can be thought of as the sum of the total angular momentum for all nucleons in a nucleus:

$$I = \sum j \qquad (3.15)$$

Since j always takes half-integer values, nuclides with an odd number of nucleons (odd A) will have odd values of nuclear spin. Similarly, nuclides with an even number of nucleons (even A) will have even values of nuclear spin. Those nuclides with odd A will have a nuclear spin wholly governed by the single, unpaired nucleon, such as occurs in ^{1}H where the single proton occupies the $1s_{1/2}$ state and $I = 1/2$. This single proton will exhibit a **nuclear magnetic moment**, μ, which is the product of the nuclear spin and an empirical constant known as the gyromagnetic ratio, γ:

$$\mu = \gamma I \qquad (3.16)$$

Alternatively, the magnetic moment of a nucleus can be expressed in terms of the nuclear g-factor, which is characteristic specific to the nucleus:

$$g = \frac{\gamma 2 m_{\mathrm{p}}}{e} \qquad (3.17)$$

where m_{p} is the mass of the proton. The existence of a nuclear magnetic moment means that nuclei will interact with an external magnetic field. Hydrogen nuclei have $I = 1/2$ and are, therefore, particularly sensitive to the effects of an external magnetic field. This forms the basis of NMR spectroscopy. Similarly, ^{13}C has a single, unpaired nucleon, this time a neutron occupying the $1p_{1/2}$ state, giving $I = 1/2$, and enabling detection through carbon-13 NMR.

Chapter Summary

- A nuclide is an atomic species which is characterized by the contents of its nucleus; isotopes have the same number of protons but a different number of neutrons; isotones have same number of neutrons but a different number of protons; isobars have the same mass number but different atomic number.
- The existence of the nucleus was first proposed by Ernst Rutherford as an interpretation of the gold-leaf experiment. The positive charge is carried by protons, while neutrons have no charge.

- The binding energy per nucleon is an important measure of nuclear stability and can be used to predict the energy output of nuclear reactions (e.g., nuclear fission).
- The liquid-drop model of nuclear structure supposes that the nucleus is an incompressible region of tightly packed nucleons which gives a constant nuclear density.
- The shell model uses a quantum mechanical approach to describing the structure of the nucleus; nucleons exist in shells defined by quantum numbers. This model can explain many experimental observations (e.g., nuclear spin).

Review Questions

(1) Classify the following as isotopes, isotones, or isobars: $^{40}_{18}$Ar and $^{40}_{19}$K; $^{14}_{6}$C and $^{16}_{8}$O; $^{40}_{19}$K and $^{40}_{20}$Ca; $^{3}_{1}$H and $^{1}_{1}$H.

(2) Free neutrons undergo beta decay and have a half-life of *ca.* 10 min. Would protons undergo a similar decay process?

(3) Calculate the binding energy per nucleon of uranium-235, stating your final answer in MeV/nucleon.

(4) Compare the nuclear radii of lithium-7 and plutonium-239.

(5) Calculate the nuclear density of an aluminium-27 nucleus and comment on how the value obtained compares with the average density of aluminium (2.70 kg·m^3).

(6) Show that the number of protons in a beryllium nucleus is consistent with the formula $n = [Z(Z-1)]/2$.

(7) Show that a plot of ln A versus ln r has a y-intercept equal to ln r_0.

(8) The compressed core of a star following a supernova explosion contains only neutrons. Calculate the density of 1 cm^3 of a neutron star, taking $r = 1.0 \times 10^{-13}$ cm and $m = 1.7 \times 10^{-24}$ g.

(9) In the semi-empirical binding energy formula, the term describing the surface effect is $-c_2 A^{2/3}$. Why is this a negative quantity?

(10) Using Eq. (3.12) and the values given in the text, calculate the binding energy per nucleon for copper-64 and zinc-64. Are these examples of isotopes, isobars, or isotones?

Chapter 4

Radioactive Decay

"Nothing in life is to be feared, it is only to be understood. Now is the time to understand more, so that we may fear less."

M. Curie

We (quite literally) live in a nuclear world — the Earth has been radioactive since its formation and continues to provide us with a range of substances, which are radioactive. On completion of this chapter and the associated questions, you should:

- Be familiar with the concept of nuclear stability and how instability leads to radioactive decay.
- Be able to identify potential alpha, beta, and gamma emitters and calculate the decay energy for the respective emission processes.
- Recognize unusual forms of radioactive decay and how they arise.
- Be able to represent decay processes graphically.

4.1 Patterns of Nuclear Stability

Key Point: Nuclear stability is governed by the ratio of protons to neutrons and the balance of repulsive and attractive forces.

In Chapter 3, we introduced the idea that nucleons in an atomic nucleus experience two of the strongest opposing forces in nature and that combination of protons and neutrons in specific shells can produce a degree of nuclear stability. To gain some idea of the relationship between the number of nucleons and nuclear stability, we can plot the number of

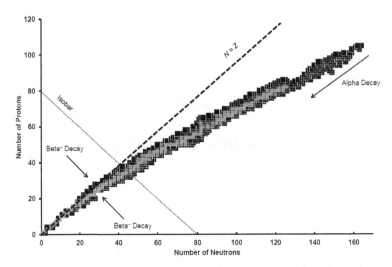

Figure 4.1 Stable and unstable nuclei and the zone of stability. The major stable nuclei (grey) and unstable nuclei (black) are plotted as N versus Z. The isobar line cuts through the plot to create a parabolic valley (see Figure 4.2).

Table 4.1 Guide to Nuclear Stability

Protons	Neutrons	Number of Stable Nuclei
Even	Even	163
Even	Odd	53
Odd	Even	50
Odd	Odd	4

protons against the number of neutrons for various stable and unstable nuclei and search for any obvious trends (Figure 4.1).

The region in grey is the so-called ***belt of stability*** and we see that all unstable isotopes keep undergoing radioactive decay until they reach this region. As atomic number increases, the deviation from the belt of stability increases, indicating that the ratio of protons-to-neutrons is decreasing. In fact, there are four ways in which the ratio of protons-to-neutrons can be defined (Table 4.1). In general, we see that the least stable nuclei are those with an odd–odd number of nucleons, such as indium-116:

$$^{116}_{49}\text{In} \rightarrow {}^{116}_{50}\text{Sn} + e^- + \bar{\nu}_e$$

The release of a beta particle (and electron anti-neutrino) brings about an increase in atomic number of one which brings it closer to the belt of stability. Conversely, when nuclei are proton-rich, stability is achieved by conversion of a proton into a neutron. This can be brought about by positron emission (e^+ emission) or through electron capture:

$$^{23}_{12}\text{Mg} \rightarrow\ ^{23}_{11}\text{Na} + e^+ + v_e$$

$$^{59}_{28}\text{Ni} + e^- \rightarrow\ ^{59}_{27}\text{Co} + v_e$$

In this case, the decrease in atomic number brings the isotopes closer to the belt of stability. Those nuclei with a large number of nucleons initially decay by alpha emission, sliding stepwise down the plot, until they are transmuted into a beta emitter, when they can finally achieve stability through either of the mechanisms described above.

In Chapter 3, we saw how the binding energy of a nucleus could provide an indication of nuclear stability. It can be shown that rewriting this equation as a function of atomic number, Z, produces a quadratic equation which defines a parabola at constant values of mass number, A:

$$BE = aA + bZ + cZ^2 \tag{4.1}$$

This parabola forms the *isobar line* in Figure 4.1 at the specified value of the mass number, cutting diagonally through the line of stable nuclei. We can visualize this "parabolic line" as a valley where the most stable nuclei lie at the bottom, while the unstable nuclei lie toward the crest. From this model, known as the *valley of stability*, we see that nuclei with lower atomic numbers are unstable with respect to β^- emission, while those with higher mass numbers are unstable with respect to β^+ emission or electron capture. In any case, after the emission event, the nuclei "slide" down the slope of the valley and occupy a position of increased stability.

4.2 Radioactive Decay

Key Point: Radioactive decay is the release of ionizing radiation from the nucleus of an atom.

Radioactive decay is generally regarded as a spontaneous nuclear transformation in which nuclei gain stability by releasing ionizing radiation. In virtually all instances, radioactive decay is independent of pressure, temperature, and chemical environment, and this allows us to model decay processes using relatively simple methods. Typically, this involves monitoring the decay of a radionuclide over a period of time and determining its half-life. Half-lives can vary from fractions of a second to millions of years, although current technologies restrict our measurements to between 10^{-18} s and 10^{15} years (*ca.* 10^5 times larger than the calculated age of the universe).

Regardless of the mode of radioactive decay, certain incontrovertible conservation laws must be observed by any radioactive decay process. We see that:

(1) The total energy of the system must be constant.
(2) The linear momentum must be conserved.
(3) The total charge must be constant.
(4) The mass number must be constant within the system.
(5) The total angular momentum of the system must be conserved.

In the language of nuclear physics, we say that a parent radionuclide undergoes decay to form *daughter nuclides*, which may or may not be radioactive themselves. The gives rise to the possibility of *decay chains*, whereby a radioactive daughter nuclide undergoes further decay to yield a new (second generation) daughter nuclide and so on.

4.3 Alpha Decay

Key Point: Alpha decay releases helium from heavy atoms, decreasing the atomic mass and increasing stability.

Alpha decay was the first recognized form of radioactive decay, initially reported by Henri Becquerel (1852–1908) and later studied by Pierre and Marie Curie. It occurs for heavier elements (typically $Z > 82$) and is characterized by the release of an alpha particle, which was shown to be a helium nucleus by Lord Rutherford in 1911. It is now our understanding that alpha particles are released from unstable nuclei as a result of Coulombic repulsion in elements heavier than lead. The general process can be represented as

$$\ce{^A_Z X} \rightarrow \ce{^{A-4}_{Z-2} X} + \ce{^4_2 He} \quad \text{e.g.,} \quad \ce{^{238}_{92} U} \rightarrow \ce{^{234}_{90} Th} + \ce{^4_2 He}$$

Alpha particles are relatively slow ($v \approx 1.5 \times 10^7 \, \text{m} \cdot \text{s}^{-1}$) with an average kinetic energy of 5 MeV; they are therefore poorly penetrating and are stopped by thin materials like paper and card.

We arrive at a value for the kinetic energy of an alpha particle through a consideration of the binding energy released on alpha decay, which corresponds to a change in mass given by the equation

$$E = mc^2 = (m_X - m_{X'} - m_\alpha)c^2 \tag{4.2}$$

where m_X is the mass of the parent radionuclide, $m_{X'}$ is the mass of the daughter nuclide, and m_α is the mass of the alpha particle, all in atomic mass units. When the masses are expressed in atomic mass units, it is convenient to express c^2 as $931.502\,\mathrm{MeV \cdot u^{-1}}$. This allows us to calculate the *decay energy* in MeV, usually referred to as the Q-value:

$$Q_\alpha = -931.502(m_X - m_{X'} - m_\alpha) \tag{4.3}$$

We regard the Q-value as the net energy released in the decay process, and provided that the decay products are in their ground states, we find that the total kinetic energy is divided between the alpha particle and the daughter nuclide:

$$Q_\alpha = KE_{\text{total}} = K_\alpha + K_{X'} \tag{4.4}$$

Due to the conservation of energy and momentum in radioactive decay, we can express the kinetic energy of the alpha particle and daughter nuclide as a function of Eqs. (4.3) and (4.4):

$$K_\alpha = \frac{Q_\alpha \cdot m_{X'}}{m_Z} \quad \text{and} \quad K_{m_{X'}} = \frac{Q_\alpha \cdot m_\alpha}{m_Z} \tag{4.5}$$

For example, the Q-value for the alpha decay of uranium-238 to thorium-234 would be given by

$$Q_\alpha = -931.502(m_X - m_{X'} - m_\alpha)$$

$$= -931.502(238.0507785 - 234.043594 - 4.002603)$$

$$= 4.274\,\mathrm{MeV}$$

The kinetic energy of the alpha particle released would therefore be

$$K_\alpha = \frac{Q_\alpha \cdot m_{X'}}{m_Z} = \frac{4.274 \times 234.043594}{238.0507785} = 4.202\,\mathrm{MeV}$$

The difference, $0.072\,\mathrm{MeV}$, will be the kinetic energy of the daughter atom, which is insignificant in comparison to the alpha particle. However, the recoil of the daughter nucleus is large in comparison to the energy of a chemical bond (*ca.* $5\,\mathrm{eV}$) which means the recoiling daughter easily breaks all associated chemical bonds.

If we compare the characteristics of a lighter nuclide undergoing alpha decay with a more typical heavier nuclide, we see a clear relationship:

$$^{144}_{60}\text{Nd} \rightarrow {}^{140}_{58}\text{Ce} + {}^{4}_{2}\text{He} \quad Q_\alpha = 1.83; \ t_{\frac{1}{2}} = 2.1 \times 10^{15}\,\text{yr}$$

$$^{262}_{107}\text{Bh} \rightarrow {}^{258}_{105}\text{Db} + {}^{4}_{2}\text{He} \quad Q_\alpha = 10.38; \ t_{\frac{1}{2}} = 4.7\,\text{ms}$$

It appears that radionuclides with short half-lives have large Q-values and *vice versa* for those with long half-lives. This suggests that the half-life of the radionuclide depends heavily on the kinetic energy of the alpha particle. This highlights an interesting conundrum which occupied physicists in the late 1920s — if alpha particles have kinetic energy between 4 and 9 MeV, and the potential energy barrier of the nucleus is around 25 MeV, how then could alpha particles be released? This problem was investigated by George Gamow (1904–1968), who in 1928 proposed the ***alpha tunneling model***,[a] in which the alpha particle is assumed to exist in a preformed state within the parent nucleus.

In Gamow's quantum mechanical treatment, it was shown that the wavefunction for the preformed alpha particle has a small, non-zero value outside the boundary of the nucleus. This implies that there is a small, yet significant probability that a preformed alpha particle can escape the potential well of the nucleus, as shown in Figure 4.2. This was one of

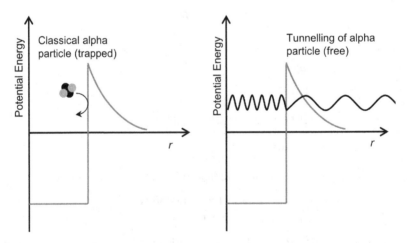

Figure 4.2 The alpha tunneling model. A classical alpha particle cannot escape the nuclear potential well, whereas the quantum mechanical model allows for a small wavefunction probability outside the potential energy barrier.

---

[a]This theory was independently proposed by Ronald Gurney (1898–1953) and Edward Condon (1902–1974) in the same year.

the first quantum mechanical triumphs of the early 20th century, and although the alpha tunneling model does not work for non-spherical nuclei, Gamow's work provided a starting point for the more advanced Nilsson–Strutinsky approach. This latter work takes into account how differently occupied nuclear shells will have varying energies, therefore affecting the nuclear potential well.

4.4 Beta Decay

Key Point: Beta decay involves the emission of electrons, positrons or the capture of electrons by light atoms in an attempt to achieve stability.

Nuclei below bismuth-209 are generally stable with respect to alpha decay (it can occur as low as $A = 106$), which implies that they achieve stability through another means, in this case beta decay. Principally, beta decay is a collective term used to refer to three processes:

(1) Electron (β^-) emission in which a down quark changes to an up quark, converting a neutron to a proton (Z increases by 1):

$$_Z^A X \rightarrow {}_{Z+1}^A X + e^- + \bar{v}_e \qquad \text{e.g.,} \quad {}_6^{14}C \rightarrow {}_7^{14}N + e^- + \bar{v}_e$$

$$Q_{\beta-} = -931.502(m_{X'} - m_X) \tag{4.6}$$

(2) Positron (β^+) emission in which an up quark changes to a down quark, converting a proton to a neutron (Z decreases by 1):

$$_Z^A X \rightarrow {}_{Z-1}^A X + e^+ + v_e \quad \text{e.g.,} \quad {}_{12}^{23}Mg \rightarrow {}_{11}^{23}Na + e^+ + v_e$$

$$Q_{\beta+} = -931.502(m_{X'} + 2M_e - m_X) \tag{4.7}$$

where the energy equivalent to the electron mass, M_e.

(3) Electron capture (EC) where a proton is converted to a neutron by combining with an electron (Z decreases by 1, A increases by 1), releasing an electron neutrino:

$$_Z^A X + e^- \rightarrow {}_{Z-1}^A X + v_e \quad \text{e.g.,} \quad {}_4^7 Be + e^- \rightarrow {}_3^7 Li + v_e$$

$$Q_{ec} = -931.502(m_{X'} - m_X) \tag{4.8}$$

Note that in the first two cases, the mass number remains constant, which means that these types of beta decay allow radionuclides to approach a more stable isobar. In electron capture, a proton-rich nucleus absorbs an inner orbital electron, transforming the nuclide into a new element. The vacant orbital is eventually filled by an outer electron, which is accompanied by emission of X-ray photons and/or Auger electrons.[b]

The passage of charged particles such as electrons and positrons out of the atom gives rise to **bremsstrahlung radiation** ("brems"). Bremsstrahlung radiation is a form of secondary radiation produced when beta particles[c] are slowed down by the electric field of the nucleus. This is often referred to as "internal bremsstrahlung," which is distinct from "outer bremsstrahlung" which arises from the movement of electrons toward the nucleus.

In beta decay, we are considering the release of two particles from the nucleus — the beta particle and a neutrino. It is possible that these particles are released in opposite directions, in which case their recoil energies cancel out and the daughter nuclide remains "stationary." However, if the beta particle and the neutrino are ejected in the same direction, the daughter nuclide will experience recoil up to a maximum of *ca.* 100 eV, in a process known as **daughter recoil**.

In the early 1920s, bismuth-210, (also known as "radium E") was studied as a model beta particle emitter:

$$^{210}_{83}\text{Bi} \rightarrow \, ^{210}_{84}\text{Po} + e^- + \bar{\nu}_e$$

If we were to apply Eq. (4.6) to the beta decay of bismuth-210, we would obtain a Q-value of 1.16 MeV. However, when the decay of bismuth-210 was investigated, a continuous energy distribution (Figure 4.3) was obtained, suggesting that beta decay does not share the same mechanism as alpha decay. If the maximum value obtained from Figure 4.3 is equal to the calculated decay energy of the beta particle, then all of the other values in the continuum must correspond to a beta particle with less energy. For this idea to observe the law of conservation of energy, Wolfgang Pauli proposed that the missing energy must be shared with another particle, which Enrico Fermi (1901–1954) later named the **neutrino**.

The inclusion of the neutrino in beta decay also mitigates problems arising from conservation of spin. In earlier chapters, we found that protons,

---

[b] An Auger electron is formed when one electron falls from a higher energy level to fill a vacancy in a lower energy level. The excess energy is released as a photon, which may knock out an orbiting electron on its exit from the atom. Electrons ejected in this way are referred to as Auger electrons.

[c] Bremsstrahlung radiation is not exclusive to beta particles; any charged particle which is slowed down sufficiently will exhibit bremsstrahlung radiation. It is a common feature of X-ray physics.

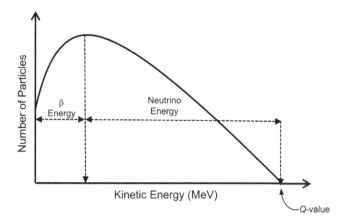

Figure 4.3 Beta decay spectrum. The continuous energy spectrum for beta decay indicates the sharing of decay energy with a further particle.

neutrons, and electrons possess spin, which may interact in various ways. The nuclear spin, I, arises from the coupling of proton and neutron spin, which have values of $s = 1/2$. Beta particles are just electrons and are therefore classified as fermions with $s = 1/2$. From Eq. (3.14) in Chapter 3, we see that the nuclear spin of bismuth-210 is 1 and that of polonium-210 is 0. This means that spin is not conserved without taking into account the spin of a neutrino, which must therefore be $s = 1/2$, i.e.,

$$^{210}_{83}\text{Bi}(I = 1) \rightarrow {}^{210}_{84}\text{Po}(I = 0) + e^- \left(I = \frac{1}{2} \right) + \bar{v}_e \left(I = \frac{1}{2} \right)$$

We also know from our earlier readings that spin can be parallel or anti-parallel. When the spin of the beta particle (electron) and neutrino are parallel we describe the process as **Gamow–Teller decay**; when the spins are anti-parallel, we speak of **Fermi decay**. Generally, the Gamow–Teller decay mode is only found in heavier nuclei and would therefore usually accompany alpha decay. Most beta decay processes follow Fermi decay.

Fermi decay is based on **Fermi's golden rule** which explains how beta decay arises and is based on the idea that the rate of decay is dependent on a quantum mechanical link between the initial and final states (i.e., parent and daughter nuclei). The link, sometimes called the coupling or matrix element, provides the probability of a decay event occurring. This is then used to define the transition probability — i.e., the probability that a beta particle is released.

The Gamow–Teller and Fermi theories of beta decay give rise to selection rules based on the change in nuclear spin. These rules describe allowed or forbidden decays. In the latter case, it is noted that certain beta decays are forbidden *per se*, but the probability of their decay is very low.

4.5 Gamma Decay

Key Point: Gamma decay is the release of gamma ray photons following alpha or beta decay, lowering the energy of excited nuclei and allowing them to achieve stability.

Gamma decay involves the release of gamma ray photons — a form of electromagnetic radiation which originates from the nucleus and has a relatively small wavelength (<10 pm) and correspondingly high energy (*ca.* 100 keV). Gamma rays are unaffected by electric and magnetic fields and are produced as a result of alpha or beta particle emission, e.g.,

$$^{226}_{88}\text{Ra} \rightarrow {}^{222}_{86}\text{Rn} + {}^{4}_{2}\text{He} + \gamma$$

$$^{234}_{90}\text{Th} \rightarrow {}^{234}_{91}\text{Pa} + e^{-} + \bar{\nu}_e + \gamma$$

The mechanism for the release of gamma ray photons can be considered to be analogous to the emission of photons of light from the movement of electrons in atomic orbitals. When a nucleus releases a beta particle, for example, the new daughter nucleus may possess excess energy above its minimum ***ground state*** energy. This ***excited state*** undergoes decay to ground state by releasing the excess energy as gamma ray photons. A common example is the decay of cobalt-60 as depicted in Figure 4.4. The release of gamma ray photons in this fashion is virtually instantaneous with the release of the alpha or beta particle ($\leq 10^{-12}$ s).

The gamma ray spectrum consists of discrete lines which is consistent with the decay processes occurring in a series of defined energy transformations. In the example shown in Figure 4.4, we see the first drop in energy produces a gamma ray photon of 1.17 MeV (*ca.* 1.87×10^{-13} J). This has a wavelength given by

$$\lambda = \frac{hc}{\Delta E} = \frac{\left(6.6 \times 10^{-34}\text{Js}\right) \times \left(1 \times 10^{7}\text{ms}^{-1}\right)}{1.87 \times 10^{-13} \text{ J}} = 3.5 \times 10^{-14}\text{m}$$

This result is in the order of size of an atomic nucleus. The second drop in energy gives a peak at 1.33 MeV, which corresponds to the final transition

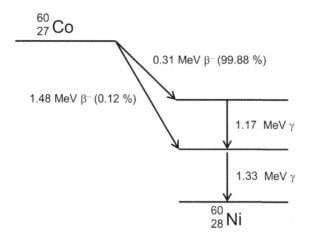

Figure 4.4 Decay scheme for cobalt-60. The initial beta decay produces an excited state, which gains stability through gamma emission.

to the ground state. Overall, the decay of cobalt-60 to nickel-60 will release one beta particle and two gamma ray photons:

$$^{60}_{27}\text{Co} \rightarrow \,^{60}_{28}\text{Ni} + e^- + \gamma(1.17\,\text{MeV}) + \gamma(1.33\,\text{MeV})$$

$$\text{or} \quad ^{60}_{27}\text{Co} \rightarrow \,^{60}_{28}\text{Ni} + e^- + 2\gamma$$

Some decay processes produce a stable excited state, referred to as a *metastable state*, which decays to its ground state through "slow" release of gamma rays. A transition from a metastable state to the ground state is known as an *isomeric transition* as the nuclide effectively remains unchanged (it is an "isomer"). One of the most common metastable nuclides is technetium-99*m*:

$$^{99m}_{43}\text{Tc} \rightarrow \,^{99}_{43}\text{Tc} + \gamma(140\,\text{keV})$$

The metastable form of technetium-99*m* has a half-life of *ca.* 6 h and on release of a gamma ray photon transforms to technetium-99 which is essentially stable with respect to further decay (half-life *ca.* 2×10^5 years).

In some nuclides undergoing alpha or beta decay, the excess energy is not always released as gamma rays, but instead through a process known as *internal conversion*. In this process, the excited nucleus interacts electromagnetically with an adjacent orbiting electron and ejects it. For example, caesium-137 initially undergoes beta decay to produce barium-137m:

$$^{137}_{55}\text{Cs} \rightarrow \, ^{137m}_{56}\text{Ba} + e^- + \bar{v}_e$$

The metastable barium-137m has a half-life of *ca.* 2.5 min and gains eventual stability by undergoing internal conversion or through gamma decay:

$$^{137m}_{56}\text{Ba} \rightarrow \, ^{137}_{56}\text{Ba}^+ + e^- \rightarrow \, ^{137}_{56}\text{Ba}$$

$$^{137m}_{56}\text{Ba} \rightarrow \, ^{137}_{56}\text{Ba} + \gamma(0.66\,\text{MeV})$$

We consider gamma decay and internal conversion as competing processes; that is, a portion of the excess energy may be released as gamma ray photons or as an electron from internal conversion. The ratio of gamma ray photons to internal conversion electrons is known as the **conversion coefficient**. Following internal conversion, the vacancy in the atomic orbital is filled rapidly by electrons from higher orbitals. This is associated with release of X-rays, which are a further form of electromagnetic radiation, although not regarded as a form of radiation.

4.6 Spontaneous Fission

Key Point: Spontaneous fission is the splitting of a heavy atom into smaller fragments; the process releases relatively large amounts of energy.

As we will discuss in Chapter 4, fission involves the splitting of a large, heavy nucleus into smaller nuclei, accompanied by the release of neutrons. For the most part, fission is an artificial process, usually induced by a flow of thermal neutrons in fission reactors. However, in the late 1930s, it was suggested that the liquid-drop model was consistent with *spontaneous fission* in superheavy elements. One year later, Russian physicists reported that uranium-238 had been observed to undergo spontaneous fission. In most instances, the large, unstable uranium-238 nucleus gains stability through alpha decay.[d] However, on some occasions, the nucleus splits apart, releasing neutrons which may be captured by adjacent non-fissioning uranium nuclei, forming uranium-239. This latter isotope is unstable with

---

[d]Beta decay is also possible in superheavy elements, but as this is mediated by the weak interaction, the rate is very slow in comparison to spontaneous fission or alpha decay. Therefore, these latter two processes are favored.

respect to beta emission and decays to neptunium-239:

$$^{238}_{92}\text{U} + ^{1}_{0}\text{n} \longrightarrow ^{239}_{92}\text{U} \xrightarrow{\beta^-} ^{239}_{93}\text{Np} + e^- + \bar{\nu}_e$$

As uranium-238 has a long half-life (*ca.* 8×10^{15} years), it undergoes significant spontaneous fission during its lifetime (for 1 kg of uranium-238 there would be around 70 fissions per second). This process is responsible for the presence of neptunium and plutonium isotopes in uranium ore. For the most part, spontaneous fission is not self-sustaining like nuclear fission as it cannot produce a steady flow of high-energy neutrons.[e] It is only energetically feasible for those nuclei with $Z^2/A \geq 235$, i.e., thorium, protactinium, uranium, and the transuranic elements.

For the heaviest nuclei, spontaneous fission is the predominant mode of radioactive decay with the likelihood of alpha decay decreasing with increasing atomic mass. The spontaneous fission of californium-252 is used in industry as a source of fast neutrons; *ca.* 3% of the nuclei decay through spontaneous fission, producing on average 3.8 neutrons per decay. Taking into account the half-life of californium-252 (2.65 year), this gives a rate of neutron production equal to 2.35×10^6 $(\mu\text{g·s})^{-1}$.

4.7 Rare Modes of Decay

Key Point: Some nuclei undergo double beta decay, proton decay, or neutron decay. These are considered rare modes of decay.

For some even–even nuclei, it is energetically favorable to undergo ***double beta decay***, e.g.,

$$^{82}_{34}\text{Se} \rightarrow ^{82}_{36}\text{Kr} + 2e^- + 2\bar{\nu}_e$$

In this process, two neutrons are converted to protons (atomic number increases by two), with the concomitant release of two neutrinos. Double EC is also possible (e.g., conversion of ^{78}Kr to ^{78}Se). Some physicists have proposed neutrinoless double beta decay in nuclides, such as germanium-76 and xenon-136. However, reports of this phenomenon have yet to be verified by independent laboratories.

---

[e]An exception to this is the Oklo natural fission reactor which sustained nuclear fission over a 200-year period. This region had very unique conditions which allowed natural ground water to mediate the flow of neutrons, controlling the fission process. It is believed that over the 200-year period, some 100 kW of energy was generated.

Proton emission can also be classified as a form of radioactive decay and is seen to be either beta-delayed or beta-prompted proton emission. It occurs in cases of extreme proton excess where a nucleus gains stability through direct emission of one or two protons. Like alpha decay, the proton can only escape the nuclear potential well through quantum mechanical tunneling. Two examples of proton emission are lithium-5 and iron-45:

$$\text{$^5_3$Li} \rightarrow \text{$^4_2$He} + \text{$^1_0$p} \quad \text{and} \quad \text{$^{45}_{26}$Fe} \rightarrow \text{$^{43}_{24}$Fe} + 2\text{$^1_0$p}$$

Similarly, *neutron emission* can also occur, which can be considered as distinct from a fission reaction as the atomic number remains the same, but the mass number decreases by 1. Therefore, both the parent and daughter nuclides are isotopes, for example,

$$\text{$^5_2$He} \rightarrow \text{$^4_2$He} + \text{$^0_1$n}$$

Neutron emission from a neutron-rich nucleus is a competing process for beta minus decay, but is much less common and is not observed in naturally occurring nuclei.

4.8 Representing Decay Processes

Key Point: Radioactive decay processes can be represented using nuclear decay schemes or decay chains. These provide more information than a simple decay equation.

So far, we have represented radioactive decay processes using radiochemical equations. While this is convenient, it does not always convey useful information, particularly with regard to mixed decay processes. Often, information on the mode of decay, the decay energy and the half-life of daughter nuclei can be included on a *nuclear decay scheme*; a simple decay scheme for the decay of cobalt-60 was shown in Figure 4.4.

We think of a decay scheme as a type of graph in which the y-axis is energy and the x-axis is proton number. For a nuclide undergoing alpha decay, an arrow moves from the parent nuclide downwards and to the left, indicating that proton number has decreased, as is the case in radon-226 (Figure 4.5). For beta decay in which a β^- particle is released, the arrow will move downwards and to the right, showing that proton number has increased; this was the case in Figure 4.4. The opposite is true for positron decay. In gamma decay processes, the arrow moves vertically from the parent to daughter nuclide as there is no change in proton number.

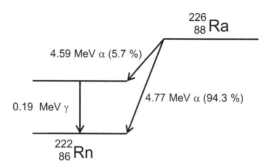

Figure 4.5 Decay scheme for radon-226. The decrease in atomic number and atomic mass is signified by arrows moving to the left-hand side.

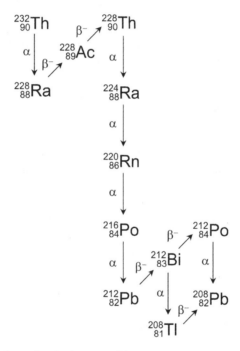

Figure 4.6 Decay scheme for thorium-232. The decay of thorium to lead occurs through 12 decay processes which constitute the "thorium series."

Another form of decay scheme used to show the entire sequence of radioactive decays is sometimes referred to as a **decay chain**. A common example of this shows the decay of throium-232 to lead-206 (Figure 4.6). In these schemes, each decay process is shown until a stable nuclide is reached. There are no axes *per se* and the direction of the arrow simply indicates progression from parent to daughter nuclei. There are four decay series, which follow the formula:

$$A = 4n + m \tag{4.9}$$

where A is the mass number of a member of the series, n is the largest integer divisible by A, and m is the remainder. From this, we obtain:

(1) The thorium series: $A = 4n$
(2) The neptunium series: $A = 4n + 1$
(3) The uranium series: $A = 4n + 2$
(4) The actinium series: $A = 4n + 3$

Only the thorium, uranium, and actinium series occur naturally; the neptunium series starts with neptunium-237 which is artificially produced. It does, however, include two naturally occurring radionuclides, bismuth-209 and thallium-205.

Chapter Summary

- Nuclear stability is largely governed by the ratio of protons to neutrons in the nucleus. Experimental evidence demonstrates that the least stable nuclei are those with an odd–odd number of nucleons.
- Nuclei gain stability through radioactive decay during which some of the mass of the nucleus is released as alpha or beta particles, with mass–energy conversion also leading to gamma ray emission.
- In alpha decay, helium nuclei, which exist pre-formed within the nucleus of the unstable atom, are released as ionizing radiation. This is explained by the alpha tunneling model.
- Beta decay can occur in three main modes: positron (β^+) emission, electron (β^-) emission or electron capture. To account for the mass difference, an additional fundamental particle, the electron (anti-) neutrino is also released.
- Gamma decay accompanies alpha or beta decay and is characterized by the release of electromagnetic radiation of low wavelength and high energy. Gamma decay is a means for nuclei in an excited state to gain stability through loss of excess energy. This sometimes involves a metastable state.
- Other forms of radioactive decay have been described, including spontaneous fission, proton emission, neutron emission, and double beta decay.

Review Questions

(1) Briefly describe (a) a plot of number of neutrons against atomic number for the stable nuclides; (b) a plot of binding energy per nucleon against mass number.

(2) Potassium has three naturally occurring isotopes: potassium-39, potassium-40, and potassium-41. Which of these isotopes is likely to be the most stable? Of those remaining, what are the likely modes of decay?

(3) Provide balanced nuclear equations for: (a) the alpha decay of thorium-230; (b) the beta decay of lead-210; the fission of uranium-235 to give barium-140, another nucleus, and two neutrons; (c) the electron capture of argon-37.

(4) Find the kinetic energy of an alpha particle emitted during the decay of radon-220, assuming that the daughter nuclide has zero recoil.

(5) Sketch a pair of Feynman diagrams for beta minus and beta plus decay (you need only to show the individual quarks involved).

(6) How much energy is released during the beta plus decay of nitrogen-13?

(7) Beryllium-8 is unusual in that it undergoes alpha decay. Provide the decay equation for this process.

(8) Free neutrons are known to undergo decay. Suggest why this is considered as the simplest form of beta decay and why a free neutron undergoes decay, yet a nucleus-bound neutron does not. Calculate the Q-value for the decay, taking the mass of a neutron as 1.008665 and the mass of a proton as 1.007825.

(9) Propose an experiment to determine if the recoil energy of beta decay is sufficient to break a C−N covalent bond.

(10) Calculate the wavelength of a 1.33 MeV gamma ray photon released during the decay of cobalt-60.

Chapter 5

Kinetics of Radioactive Decay

"Change isn't easy, it takes time."

C. Kennedy

Since the discovery of radioactivity in 1896 by Henri Becquerel, what was once regarded as an almost magical phenomenon has since been shown to obey relatively simple laws. On completion of this chapter and the associated questions you should:

- Be able to use appropriate units to describe radioactive decay.
- Be able to derive and apply the universal law of radioactive decay.
- Have an understanding of more complex decay kinetics.
- Appreciate the origins of uncertainty in radioactive decay.

5.1 Measuring Radioactive Decay

Key Point: The SI unit of radioactivity is the becquerel, Bq, which corresponds to one disintegration per second.

When an unstable atom undergoes decay, the ionizing radiation produced can be measured by a variety of scientific instruments. More specific details regarding the physical measurement of radioactive decay will be given in Chapter 7. For now, we will be concerned with the data gathered from such measurements. When the sample decays, ionizing radiation is released in all directions, suggesting that only a portion of the radiation will interact with the detector. When radioactivity is measured in this way, we can only deal with the ***observed decay rate***, R, which is typically measured in **counts per minute**, cpm. This can be related to the **activity**, A (the "rate of decay") which is measured in disintegrations per second; one disintegration per second is known as a **becquerel**, Bq, and this is the SI unit of radioactive decay. An older non-SI unit of radioactivity still in use is the **curie**, Ci, which is related to the becquerel:

$$1\,\text{Ci} = 3.7 \times 10^{10}\,\text{Bq} = 37\,\text{GBq}$$

The **specific activity**, S, of a radioisotope is defined as the activity per unit mass of the element/compound. In SI units this would have dimensions of Bq/kg:

$$S = \frac{A(\text{Bq})}{m(\text{kg})} \tag{5.1}$$

There are a number of other units associated with radioactivity which will be covered in Chapter 7.

5.2 The Universal Law of Radioactive Decay

Key Point: Radioactive decay obeys simple first-order kinetics; a key feature of radioactive decay is the half-life of a radionuclide.

In order to model the behavior of radionuclides, we look to the familiar laws of mathematics and chemical kinetics, namely those governing exponential processes. From basic chemistry, we know that the rate of reaction is dependent upon the amount of reactant present; for some reactions, if the concentration of reactant is doubled, the rate doubles in what is often referred to as a first-order reaction. From experimental evidence, it appears that radioactive decay can also be modeled by first-order kinetics; however, great care must be taken not to confuse a chemical reaction with a nuclear reaction — the usual factors affecting the rate of a chemical reaction (temperature, pressure etc.) have no effect on nuclear processes. In this sense, we are simply borrowing a model from chemical kinetics and applying it to another physical process. In fact, a similar approach can also be used to model population growth, which clearly has no link to chemical processes.

First, let us consider a 1 g sample of radium-223, which can be shown to contain 2.7×10^{21} radium nuclei:

$$\frac{1\,\text{g}}{223\,\text{g/mol}} \times 6.022 \times 10^{23}\,\text{particles/mol} = 2.7 \times 10^{21}\,\text{particles}$$

This means that in 1 g of radium-223, there are well over a trillion nuclei all capable of undergoing alpha particle emission at any given instant.

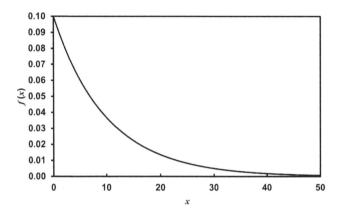

Figure 5.1 Exponential distribution. The exponential function, $f(x) = e^{-x}$, shows a constant decrease with increasing values of x.

Obviously, this number is so large that it would be impossible to try and build a model with the exact number of nuclei present. Instead, we make the assumption that because the number of nuclei is so large, they will behave as a continuous variable. Based on this assumption, we can use the mathematics describing continuous variables to build our model of radioactive decay, in this case, the *exponential distribution*. We can visualize the exponential distribution as shown in Figure 5.1 where we see it is characterized by a steady decline in the probability of an event occurring with time.

Suppose we measured the alpha particle emission from 1 g sample of radium-223. If we were to plot the results, alpha particle count versus time, we would obtain a graph similar to that shown in Figure 5.2. We immediately see that the data represented in this figure follows an exponential distribution, leading us to the conclusion that our original assumption about the number of nuclei behaving as a continuous variable was correct. With this in mind, let us say that the number of nuclei present is represented by N. With the passage of time, t, the rate of radioactive decay (the activity, A) will decrease as the number of nuclei decrease. Using the calculus notation introduced in Chapter 1, we say that the rate of decay is proportional to N:

$$A = -\frac{dN}{dt} \quad \text{and} \quad -\frac{dN}{dt} \propto N \qquad (5.2)$$

This implies that $A \propto N$, which is entirely sensible; the greater the number of nuclei present, the greater the activity detected.

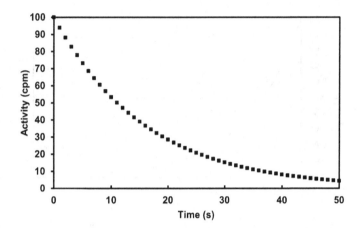

Figure 5.2 Decay curve for radium-223. As time progresses, the number of nuclei declines in an exponential fashion.

We introduce the **decay constant**, λ, into Eq. (5.2), which can be thought of as analogous to the rate constant in a first-order rate equation:

$$-\frac{\mathrm{d}N}{\mathrm{d}t} = \lambda N \tag{5.3}$$

Equation (5.3) is a differential equation, which states that the rate of decay, $-\mathrm{d}N/\mathrm{d}t$, is equal to the original number of nuclei present multiplied by the decay constant; or, written another way:

$$A = \lambda N \tag{5.4}$$

It is important to realize that the decay constant is not a true constant; its value varies with each radionuclide under study. For radium-223, the decay constant has a value of 0.0607/day which means the radium-223 nuclide will release, on average, an alpha particle every 0.0607 of a day (*ca.* one every 1.5 h).

If we wish to use Eq. (5.3) in any meaningful way, we must solve it by integration. Firstly, we rearrange the equation to collect together all the "N-terms" on the left-hand side and place the integral signs. To make future steps more straightforward, we also move the negative sign to the right-hand side:

$$\int \frac{\mathrm{d}N}{N} = -\lambda \int \mathrm{d}t \tag{5.5}$$

The next stage is to define the limits of our integration. At the very start, $t = 0$ and there will be an initial number of nuclei present, N_0. At some time after this, designated $t = t_1$, there will be fewer nuclei present, designated N_t. We include these limits in our integration:

$$\int_{N_0}^{N_t} \frac{dN}{N} = -\lambda \int_{t_0}^{t_1} dt \tag{5.6}$$

We can now complete the integration using the rules from Appendix A, taking care to note that $dN/N = 1/N \cdot dN$. This gives

$$(\ln N_t - \ln N_0) = -\lambda(t_1 - t_0) \tag{5.7}$$

Next, we want to remove the natural logarithms by using the exponential constant:

$$N_t - N_0 = e^{-\lambda(t_1 - t_0)} \tag{5.8}$$

At this point, we notice that the term $(t_1 - t_0) = t_1$ because $t_0 = 0$. Therefore, Eq. (5.8) can be simplified to what is known as the **universal law of radioactive decay**:

$$N_t = N_0 e^{-\lambda t} \quad \text{or} \quad A_t = A_0 e^{-\lambda t} \tag{5.9}$$

Consider a 0.5 g sample of plutonium-239 which has a decay constant of $9.1 \times 10^{-13}\,\text{s}^{-1}$. Suppose we wanted to calculate the activity of the plutonium after 17 years. First, we calculate the initial activity by determining the number of nuclei present:

$$\frac{0.5\text{g}}{239\text{g/mol}} \times 6.022 \times 10^{23} \text{ nuclei/mol} = 1.3 \times 10^{21} \text{ nuclei}$$

The initial activity will be this number multiplied by the decay constant:

$$A = \lambda N = (9.1 \times 10^{-13}\text{s}^{-1}) \times (1.3 \times 10^{21} \text{ nuclei}) = 1.2 \times 10^9 \text{ Bq}$$

Finally, the activity remaining after 17 years can be calculated from Eq. (5.9):

$$A_t = 1.2 \times 10^9\text{Bq} \times \{\exp-[(9.1 \times 10^{-13}\,\text{s}^{-1}) \times (17\,\text{yr} \times 3.1 \times 10^7\text{yr} \cdot \text{s}^{-1})]\}$$
$$= 1.2 \times 10^9 \text{ Bq}$$

We see in this example that the activity has not decreased within the significant figures used. This is because plutonium-239 has such a long half-life (*ca.* 24000 years) that the exponential term is virtually zero.

Now we will relate half-life to the universal law of radioactive decay. We have previously defined the half-life as the time taken for the activity to decrease to half its original value. As activity and number of nuclides will be proportional, we can also consider half-life to be the time taken for the original number of radionuclides to decrease by half. This means that when $t = t_{1/2}$, $N_t = N_0/2$, which substituted into Eq. (5.9) gives

$$\frac{N_0}{2} = N_0 e^{-\lambda t_{1/2}} \tag{5.10}$$

Dividing both sides by N_0 and taking natural logarithms gives

$$\frac{1}{2} = e^{-\lambda t_{1/2}} \Rightarrow \ln\frac{1}{2} = -\lambda t_{1/2} \tag{5.11}$$

Using the rules of logs we find that $\ln(1/2) = \ln(1) - \ln(2) = 0 - \ln(2)$, i.e.,

$$t_{1/2} = \frac{\ln 2}{\lambda} \quad \text{or} \quad t_{1/2} = \frac{0.693}{\lambda} \tag{5.12}$$

Thus, if we know the decay constant for a particular radioisotope, we can determine the half-life relatively accurately and *vice versa*.

One useful approach to determining the value of the decay constant is to construct a plot of ln (activity) versus time, such as that shown in Figure 5.3 for carbon-11. Recalling that we can treat "number of nuclides" and "activity" in the same way, we see that this plot follows Eq. (5.7) written in the form:

$$\ln A_t = \ln A_o - \lambda t \tag{5.13}$$

This follows the equation of a line ($y = c - mx$) which means the gradient is equal to the decay constant. For Figure 5.3, the linear regression equation is

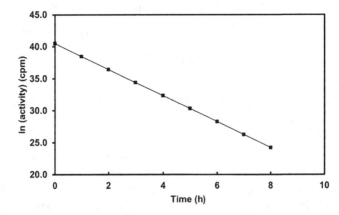

Figure 5.3 Plot of $\ln(A)$ versus time for carbon-11. The gradient of the regression line is equal to the decay constant with dimensions of reciprocal time.

$y = -2.037x + 40.508$ which gives us $\lambda = 2.037\,\mathrm{h}^{-1}$. Then, using Eq. (5.12), this gives us a half-life for carbon-11 of 0.34 h or 20.4 min.

A useful consequence of Eqs. (5.9)–(5.12) is that if we know the original number of nuclides and the number of half-lives, we can directly evaluate the number of nuclides remaining after n half-lives:

$$N = \frac{N_0}{2^n} \qquad (5.14)$$

For example, if we had $3.5\,\mu\mathrm{g}$ of carbon-11 we can easily determine that this would contain 1.92×10^{17} nuclei. After three half-lives there would be

$$N = \frac{1.92 \times 10^{17}}{2^3} = 2.4 \times 10^{16} \text{ nuclei}$$

5.3 Decay in a Mixture of Isotopes

Key Point: The total activity in a mixture of isotopes is the sum of the individual activities.

Often when working with radioisotopes in practical situations, we may have a mixture of two or more independent isotopes — e.g., sodium-24 and phosphorous-32 — which have the same mode of decay but with different half-lives. It is possible that, depending upon the age of the sample, the observed radioactivity is due to contributions from both components. The total observed activity would be the sum of the individual activities:

$$A = A_1^0 e^{-\lambda_1 t} + A_2^0 e^{-\lambda_2 t} \qquad (5.15)$$

where A_1^0 and λ_1 signify the initial activity and decay constant for one of the isotopes in the mixture. If we multiply through by $e^{-\lambda_1 t}$, we obtain an equation which produces a linear plot of $Ae^{-\lambda_1 t}$ versus $e^{t(\lambda_1 - \lambda_2)}$ with gradient of A_2^0 and y-intercept A_1^0:

$$Ae^{-\lambda_1 t} = A_1^0 + A_2^0 e^{t(\lambda_1 - \lambda_2)} \qquad (5.16)$$

An example of this is shown in Figure 5.4 for the decay of a mixture of silver-108 ($t_{1/2} = 142.3\,\mathrm{s}$) and silver-110 ($t_{1/2} = 24.6\,\mathrm{s}$). The linear regression equation is $y = 0.0011x + 0.0006$ which gives the initial activity of silver-108 as $0.0011\,\mathrm{MBq}$ and silver-110 as $0.0006\,\mathrm{MBq}$.

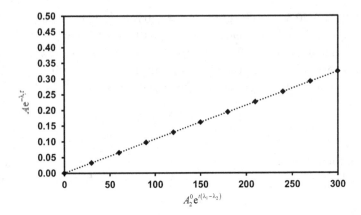

Figure 5.4 Decay of a mixture of silver-108 and silver-110. The gradient of the line gives the initial activity of silver-110 and the y-intercept gives the initial activity of silver-108.

5.4 Branched Decay Kinetics

Key Point: The simultaneous decay of a parent nuclide into two or more daughter nuclides is characterized by partial half-lives.

We have already encountered many instances of **branched decay** in which a parent radionuclide undergoes simultaneous decay processes. For example, copper-64 has a half-life of *ca.* 12.7 h and decays by electron capture (43%) and β^+ emission (19%) to nickel-64, and by β^- emission (38%) to zinc-64. We can characterize each of these modes of decay with **partial decay constants** (λ_1, λ_2, and λ_3) which are related to the overall decay constant:

$$\lambda = \lambda_1 + \lambda_2 + \lambda_3 \tag{5.17}$$

Similarly, the **partial half-lives** are related to the overall half-life of the parent nuclide:

$$\frac{1}{t_{1/2}} = \frac{1}{t_{1/2}^1} + \frac{1}{t_{1/2}^2} + \frac{1}{t_{1/2}^3} \tag{5.18}$$

In the case of copper-64 ($t_{1/2} = 12.6$ h; $\lambda = 0.055$ h^{-1}), we can evaluate each of the individual decay constants and half-lives relatively easily:

$$\lambda_1 = 0.055 \times 0.43 = 0.024 \, \text{h}^{-1} \quad \text{and} \quad t_{1/2}^1 = 28.9 \, \text{h}$$

$$\lambda_2 = 0.055 \times 0.19 = 0.010 \, \text{h}^{-1} \quad \text{and} \quad t_{1/2}^2 = 69.3 \, \text{h}$$

$$\lambda_3 = 0.055 \times 0.38 = 0.021 \, \text{h}^{-1} \quad \text{and} \quad t_{1/2}^3 = 33.0 \, \text{h}$$

We can easily verify this as the sum of the partial decay constants returns the overall decay constant.

5.5 Successive Decay Kinetics

Key Point: The total activity of a decay chain depends on the half-lives of parent and daughter nuclei; often, equilibrium is established.

When an unstable radionuclide undergoes decay its daughter nuclides often undergo further decays. For example, caesium-139 decays according to the scheme:

$$^{139}_{55}\text{Cs} \xrightarrow[9.5\,\text{min}]{\beta^-} {}^{139}_{56}\text{Ba} \xrightarrow[82.9\,\text{min}]{\beta^-} {}^{139}_{57}\text{La}$$

The decay of caesium-139 will obey Eq. (5.9) *et seq.*; however, the decay of barium-139 is more complex as its decay will be a function of its formation from caesium-139 and its decay to lanthanum-139. We will assume that we start with N_0 atoms of the parent at $t = 0$ and that no daughter atoms are present. Therefore, we have

$$N_1(t = 0) = N_0 \quad \text{and} \quad N_2(t = 0) = N_3(t = 0) \text{ etc.}$$

It follows that the decay of barium-139 will be the difference in the rates of these two processes:

$$\frac{\mathrm{d}N_2}{\mathrm{d}t} = \lambda_1 N_1 - \lambda_2 N_2 \tag{5.19}$$

We already have an expression for N_1, it is simply Eq. (5.9), so this can be substituted in Eq. (5.19) to give

$$\frac{\mathrm{d}N_2}{\mathrm{d}t} = \lambda_1 N_0 \mathrm{e}^{-\lambda_1 t} - \lambda_2 N_2 \tag{5.20}$$

Equation (5.20) is an example of a linear differential equation with constant coefficients and solving it is a challenging exercise in algebra and calculus. Instead, we will assume that the correct solution to this equation for the initial condition $N_2(0) = 0$ is

$$N_2 = \frac{\lambda_1}{\lambda_2 - \lambda_1} N_0 (\mathrm{e}^{-\lambda_1 t} - \mathrm{e}^{-\lambda_2 t}) \tag{5.21}$$

Equivalently, we can express Eq. (5.21) in terms of the activity of the daughter radionuclide by multiplying through by λ_2:

$$\lambda_2 N_2 = \frac{\lambda_2 \lambda_1}{\lambda_2 - \lambda_1} N_0 \left(\mathrm{e}^{-\lambda_1 t} - \mathrm{e}^{-\lambda_2 t} \right) \tag{5.22}$$

Taking into account that the activity of the parent at time t is $\lambda_1 N_1 = \lambda_1 N_0 e^{-\lambda_1 t}$ we can divide Eq. (5.22) by this to get the ratio of daughter-to-parent activity:

$$\frac{\lambda_2 N_2}{\lambda_1 N_1} = \frac{\lambda_2}{\lambda_2 - \lambda_1}[1 - e^{-(\lambda_2 - \lambda_1)t}] \tag{5.23}$$

5.5.1 Equilibrium in parent–daughter activities

In a decay series, there will be a point at which the rate of decay of a daughter nuclide is equal to the rate of its production (if the parent has only one daughter). At this point, the rate of change in the daughter's activity will be zero:

$$\frac{d(\lambda_2 N_2)}{dt} = (-\lambda_1 e^{-\lambda_1 t} + \lambda_2 e^{-\lambda_2 t}) = 0 \tag{5.24}$$

If we rearrange for time, we will arrive at an expression for the time at maximum activity, t_{max}:

$$t_{max} = \frac{1}{\lambda_2 - \lambda_1} \cdot \ln\left(\frac{\lambda_2}{\lambda_1}\right) \tag{5.25}$$

The specific relationship between the daughter and parent activity depends on the relative size of the decay constants/half-lives. This gives rise to three general possibilities: transient equilibrium, secular equilibrium, and non-equilibrium.

5.5.1.1 *Transient equilibrium*

When the half-life of the parent nuclide is longer than that of the daughter, we will have $\lambda_2 > \lambda_1$. Assuming that the counting efficiency is the same for both radionuclides, we would initially observe an increase in the total activity. However, with increasing time, the exponential term in Eq. (5.23) becomes smaller and the ratio approaches a constant value:

$$\frac{\lambda_2 N_2}{\lambda_1 N_1} = \frac{\lambda_2}{\lambda_2 - \lambda_1} = \text{constant} \tag{5.26}$$

The maximum activity of the daughter will be given by Eq. (5.25). After this point, the parent and daughter decay at virtually the same rate, establishing a transient equilibrium.

A common example of a transient equilibrium is that between molybdenum-99 and technetium-99m:

$$^{99}_{42}\text{Mo} \xrightarrow[66.7\,\text{h}]{\beta^-} {}^{99m}_{43}\text{Tc} \xrightarrow[6.03\,\text{h}]{\gamma} {}^{99}_{43}\text{Tc}$$

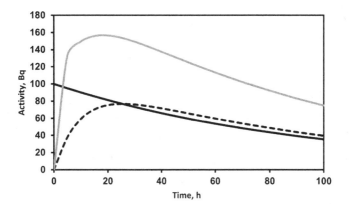

Figure 5.5 Transient equilibrium. Molybdenum-99 decays smoothly (black line) while technetium-99m (dashed line) increases to a maximum and then declines. The total activity (grey line) reaches a maximum slightly before $t_{\max}$, but then declines.

Molybdenum-99 decays to technetium-99m in 86% of cases; in the remaining 14% it decays directly to technetium-99. Therefore, the decay constant, λ_1 describes a branched decay (*c.f.* Section 5.4) and the decay constant associated with formation of technetium-99m will be $(0.693/66.7) \times 0.86 = 0.0089\,\text{h}^{-1}$. Taking this into account, we can plot the behavior of the radionuclides (Figure 5.5) to show the transient equilibrium. The time at which the technetium-99m activity is at a maximum and is given by Eq. (5.25):

$$t_{\max} = \frac{1}{0.115 - 0.0104} \cdot \ln\left(\frac{0.115}{0.0104}\right) = 23.0\,\text{h}$$

The advantage of transient equilibrium is that the parent isotope can be prepared and delivered to its point of use without any negative effect on the total activity of the desired daughter radionuclide.

It would be instructive to see how Eq. (5.26) can be used. Suppose pure molybdenum-99 was prepared and shipped to a laboratory with an initial activity of 11.1 GBq. The activity of the technetium-99m some period of time later, say 6 h, can be obtained from the ratio of activities given by Eq. (5.26) (remembering to take λ_1 as $0.0089\,\text{h}^{-1}$ to take into account the mixed decay). This gives

$$\frac{\lambda_2 N_2}{\lambda_1 N_1} = \frac{0.115}{0.115 - 0.0089} = 1.084$$

Since we know that $\lambda_1 N_1 = 11.1\,\text{GBq}$ we can determine $\lambda_2 N_2$ at $t = 6\,\text{h}$ using Eq. (5.23):

$$\lambda_2 N_2 = 11.1 \times 1.084 \times \{1 - [\exp(-0.115 - 0.0089) \times 6]\} = 5.6\,\text{GBq}$$

We have already shown that the maximum activity of technetium-99m will be at 23 h; if we let $t = 23$ in Eq. (5.23) we will therefore get the maximum possible activity for technetium-99m from the original 11.1 GBq of molybdenium-99:

$$\lambda_2 N_2 = 11.1 \times 1.084 \times \{1 - [\exp(-0.115 - 0.0089) \times 23]\} = 10.98 \ \text{GBq}$$

We see that this is 98% of the molybdenum-99 activity, which can provide a useful approximation of the activity at $t_{\max}$.

5.5.1.2 *Secular equilibrium*

When $\lambda_2 \gg \lambda_1$, the half-life of the parent nuclide is much longer than that of the daughter and it effectively decays at a constant rate. Although the process still follows Eq. (5.23), for all practical purposes the exponential term is ~ 1 and we see that it becomes

$$\frac{\lambda_2 N_2}{\lambda_1 N_1} = \frac{\lambda_1}{\lambda_2 - \lambda_1} \simeq 1 \tag{5.27}$$

This implies that initially the activity of the sample will be dominated by the parent radioisotope (when $t = 0$, $N_2 = 0$). Then, within about seven daughter half-lives, the activities are equal and the total activity appears to have doubled. Beyond this, the activity effectively remains at a constant level for an almost indefinite amount of time in what is referred to as a secular equilibrium. In fact, this is really a special case of transient equilibrium.

An example of a secular equilibrium is shown in Figure 5.6 for the decay of radium-226 to radon-222 which is part of the uranium-238 decay series:

$$^{226}_{88}\text{Ra} \xrightarrow[1602\,\text{yr}]{\alpha,\gamma} {}^{222}_{86}\text{Rn} \xrightarrow[3.824\,\text{d}]{\alpha} {}^{218}_{84}\text{Po}$$

From Eq. (5.25), the time of maximum activity would be 66 days, but as the maximum activity is virtually constant, this value really represents the onset of maximum activity. At this point, the ratio of activities would be almost unity (this is not clear in Figure 5.6 — the daughter growth curve approaches the parent curve almost asymptotically). The advantage of a secular equilibrium is that it allows you to predict the activity of the daughter radioisotope. For example, if we had 10 MBq

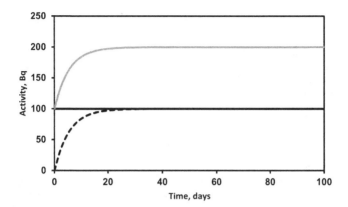

Figure 5.6 Secular equilibrium. Radium-226 decays almost imperceptibly (black line) while radon-222 (dashed line) increases to a maximum and remains at this level. The total activity (grey line) mirrors the pattern of daughter activity, except at double the activity level.

of radium-226 initially, then after 66 days, there would be 10 MBq of radon-222.

5.5.1.3 *Non-equilibrium*

If the parent nuclide has a shorter half-life than the daughter, $\lambda_1 > \lambda_2$, the parent will decay quickly and the daughter activity will rapidly rise to a maximum given by the reverse of Eq. (5.23):

$$\frac{\lambda_2 N_2}{\lambda_1 N_1} = \frac{\lambda_1}{\lambda_1 - \lambda_2} \left(e^{(\lambda_1 - \lambda_2)t} - 1 \right) \tag{5.28}$$

In this case, it should be obvious that no equilibrium can occur and the overall activity would appear to decline. The t_{max} can still be evaluated from Eq. (5.25) even though the equilibrium condition is not satisfied. If the parent decays to a stable daughter, then the number of daughter nuclei will exactly equal the number of parent nuclei that have undergone decay. This scenario is relatively uncommon but is seen in a few radionuclides such as tellurium-131m (Figure 5.7):

$$^{131m}_{52}\text{Te} \xrightarrow[30\,\text{h}]{\beta^-} {}^{131}_{53}\text{I} \xrightarrow[193\,\text{h}]{\beta^-} {}^{131}_{54}\text{Xe}$$

As before, we see that the daughter activity increases up until t_{max} (*ca.* 95 min in this example). However, this has no effect on the total activity, which steadily declines from the start of measurement.

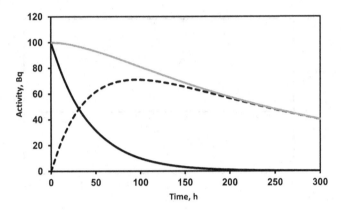

Figure 5.7 Non-equilibrium. Tellurium-131*m* Radium-226 decays smoothly (black line) while iodine-131 (dashed line) increases to a maximum and then decays. The total activity (grey line) decays relatively smoothly throughout.

Chapter Summary

- Radioactivity is practically measured in counts per second; this is converted into disintegrations per second by taking into account the counter efficiency. One disintegration per second is known as one becquerel of activity.
- Radioactive decay follows simple first-order kinetics with the activity of a nuclide being proportional to the number of nuclei present ($A \propto N$). The constant of proportionality is known as the decay constant ($A = \lambda N$) which is related to the half-life of the radionuclide ($t_{1/2} = \ln 2/\lambda$). These features arise from the universal law of radioactive decay.
- More complex kinetics are required to describe mixtures of isotopes, decay chains and branched decays. These are all based on the universal law of radioactive decay.
- Radioactive is a fundamentally quantum mechanical process and is subject to an inherent degree of uncertainty. We can use the lifetime of a nuclide to determine the uncertainty associated with a particular nuclear transformation.

Review Questions

(1) What mass (in g) of iodine-131 is equivalent to 37 MBq, taking $t_{1/2} = 8.02$ days?

(2) Gadolinium-152 decays by alpha emission and is present in the naturally occurring element at an abundance of 0.20%. If a 5.0 g sample of metallic gadolinium was found to release 114 alpha particles every 4 h, what is the half-life of the radioisotope?

(3) Naturally occurring potassium contains 0.012% potassium-40 ($t_{1/2} = 1.3 \times 10^9$ years). Assuming that the human body contains 0.35% potassium by weight, calculate the total radioactivity resulting from potassium-40 in an average 75 kg human.

(4) What is the theoretical maximum specific activity (in $\text{MBq} \cdot \text{mmol}^{-1}$) that can be obtained in a sample of [^{32}P]fructose-1,6-diphosphate, taking $t_{1/2} = 14.3$ days?

(5) Carbon-14 decays with a half-life of 5730 years. If an object originally containing 8.50 μg of carbon-14 was found to contain 0.80 μg, what is the age of the sample?

(6) If 5.0 μg of calcium-47 ($t_{1/2} = 4.5$ days) are required for an experiment, what mass of ^{47}CaCO$_3$ should be requested from the supplier, taking into account that it takes 48 h for delivery?

(7) A 0.1 mL bolus of a radiotracer ($5 \times 10^3 \, \text{cps} \cdot \text{mL}^{-1}$) was injected into a laboratory rat. Several minutes later, a 1 mL blood sample was drawn and the measured radioactivity was 48 cpm. Estimate the total blood volume of the rat.

(8) Yttrium-87 ($t_{1/2} = 80$ h) is in a transient equilibrium with its daughter radionuclide strontium-87m ($t_{1/2} = 2.83$ h). If 11.1 GBq of yttrium-87 is delivered to a radiopharmacy department 6 h after it is synthesized, calculate the activity of the strontium isotope.

(9) Beryllium-8 has $t_{1/2} = 0.07$ fs. How long will it take for a 2.0 μg sample to decay to 0.1% of its original activity?

(10) Polonium-210 ($t_{1/2} = 138$ days) decays to lead-206 through alpha emission. What volume of helium at 25°C and 1 atm would be obtained from a 25 g sample of polonium-210 after one half-life?

Chapter 6

Nuclear Reactions

"People must understand that science is inherently neither a potential for good nor for evil. It is a potential to be harnessed by man to do his bidding."

G. Seaborg

For many people, the idea of a nuclear reaction conjures images of mushroom clouds following a thermonuclear explosion, or the devastation left after the Chernobyl incident. In fact, nuclear reactions are occurring in our upper atmosphere every second, as well as in our surrounding environment. On completion of this chapter and the associated questions you should:

- Be able to calculate the reaction cross-section for simple nuclear reactions involving thermal neutrons.
- Relate models of nuclear structure to the main types of nuclear reaction (fission and fusion).
- Have an understanding of the main routes for the preparation of nuclear materials through artificial nuclear reactions.

6.1 Overview of Nuclear Reactions

Key Point: Nuclear reactions occur between nuclei and energetic particles to produce new, radioactive nuclei.

Nuclear reactions involve the transformation of one isotope of an element into another, or its conversion into an entirely different element. The initial products of a nuclear reaction are unstable with respect to radioactive decay and they are therefore regarded as radionuclides. Distinct from these artificial radionuclides are those which occur naturally. The **primordial radionuclides** are those which were formed through stellar nucleosynthesis and include isotopes of uranium and thorium. Secondary radionuclides are formed through the radioactive decay of primordial nuclei, such as the formation of radium from uranium. Cosmogenic radionuclides are created by the action of cosmic rays on the Earth's atmosphere; chief among these is nitrogen-14 which decays to carbon-14 and forms the basis of radiocarbon dating.

In 1917, Ernst Rutherford performed the first artificial nuclear reaction by converting nitrogen atoms into a stable isotope of oxygen through bombardment with alpha particles. We can represent nuclear reactions such as this in much the same way as chemical equations or nuclear decay processes. For example, Rutherford's reaction can be represented as

$$^{14}_{7}N + \alpha \rightarrow {}^{17}_{8}O + p$$

A more succinct notation is $^{14}N(\alpha,p)^{17}O$ where the terms in brackets relate to the particles absorbed and released in the process.

In 1934, Frédéric Joliot-Curie (1900–1958) and Irene Joliot-Curie (1897–1956) transformed aluminium to phosphorous-30 through alpha particle bombardment, $^{12}Al(\alpha,n)^{30}P$. If we consider this reaction, we should realize that in order for an alpha particle to penetrate the Coulomb barrier of the aluminium nucleus, it must have high kinetic energy. In this case, the alpha particles were produced by polonium-210 decay and accelerated by passing them through a high-voltage ion accelerator. Phosphorus-30 is unstable ($t_{1/2} = 2.5$ min) and the Joliot-Curies were awarded The Nobel Prize in Chemistry for successfully inducing artificial radioactivity in a substance.

Of course, most nuclear reactions are more complex and can involve a series of steps. A few years after Chadwick's discovery of the neutron, Enrico Fermi and colleagues started experiments bombarding uranium with neutrons. Fermi originally believed they had created two new elements, but this was met with skepticism and it was not until Otto Hahn (1879–1968), Lise Meitner (1878–1968), and Fritz Strassman (1902–1980) recreated this work that they discovered nuclear fission — the splitting of a heavy element into smaller fragments. This event, dubbed the "splitting of the atom," marked the start of the nuclear age, from which we have developed nuclear power and a number of technologies which have allowed us to further our understanding of nuclear structure.

6.2 Likelihood of Nuclear Reactions

Key Point: The likelihood of a successful nuclear reaction is governed by the reaction cross-section, measured in barn.

We have already noted that a relatively small mass of material will contain many individual nuclei. It follows that in terms of nuclear reactions, we must take into account the likelihood of a target nucleus being struck by an incoming particle. The **number density** of target nuclei, n, is easily evaluated from the physical dimensions of the target:

$$n(\text{cm}^{-3}) = \frac{\rho(\text{g/cm}^3) \cdot N_A(\text{atoms/mol})}{A(\text{g/mol})} \tag{6.1}$$

where ρ is the density of the target, N_A is Avogadro's number and A is the atomic mass of the target. Given the order of the nuclear scale, number densities are large numbers, up to *ca.* $10^{20}\,\text{cm}^{-3}$. If a beam of particles is directed at the target, not all particles will interact with the target nuclei and we see that the likelihood of a successful collision depends on the ratio of the effective area of the target nucleus to the overall area of the target. This ratio is known as the **reaction cross-section**, σ (sigma), which is measured in units of *barn*, b, where $1\text{b} = 1 \times 10^{-28}\,\text{m}^2$ (or $1 \times 10^{-24}\,\text{cm}^2$).[a]

If we assume that a reaction will only occur if the incident particle strikes the reaction cross-section, we can propose that the probability of a successful collision is directly proportional to the reaction cross-section. Consequently, the total area exposed to the incident particles must be a function of this, *viz.* $\sigma n x$ where x is the volume of the target material. If we represent the incident particle flux as φ (phi), we arrive at the scenario depicted in Figure 6.1. The target is exposed to a homogeneous flux of particles and a portion of these are absorbed in the conversion process. The change in particle flux can be expressed as a function of the thickness of the target, x:

$$-\frac{\mathrm{d}\phi}{\mathrm{d}x} = \phi \sigma n \tag{6.2}$$

---

[a]This unit was introduced in 1942 by M.G. Holloway and C.P. Baker who described the value $1 \times 10^{-28}\,\text{m}^2$ as being "as big as the broad side of a barn" for nuclear processes. One barn is approximately equal to $\pi(6 \times 10^{-15})^2$ where $6 \times 10^{-15}\,\text{m}$ is the average nuclear radius.

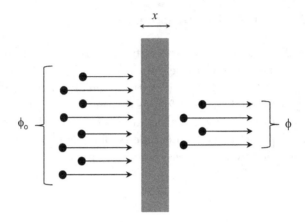

Figure 6.1 Attenuation of energetic particles. When a flow of energetic particles is directed at a target, a portion of them is absorbed, which is dependent on the thickness and cross-section of the material.

Integrating in the usual way by setting $\varphi = \varphi_0$ when $x = 0$ gives

$$\int_{\phi_0}^{\phi} \frac{\mathrm{d}\phi}{\phi} = -\sigma n \int_0^x \mathrm{d}x \Rightarrow \ln\left(\frac{\phi}{\phi_0}\right) = -\sigma n x$$

$$\therefore \ \phi = \phi_0 \mathrm{e}^{-\sigma n x} \tag{6.3}$$

It follows that by taking appropriate measurements of φ, the reaction cross-section can be determined for a particular target.

6.3 Formation of Radionuclides

Key Point: Radionuclides are produced by directing energetic particles at a target material in either a nuclear reactor or a particle accelerator.

In modern-day practice, radionuclides are formed using either nuclear reactors or particle accelerators. Reactor-produced radionuclides are neutron-rich and tend to decay through beta emission, while those produced in particle accelerators are neutron-deficient and decay through release of positrons. In both cases, generating radionuclides involves directing a flow of energetic particles at a ***target***. The nature of the target depends on whether a nuclear reactor or particle accelerator is being used (the same basic theory applies to nuclear power stations, but this discussed in Chapter 12). Targets used in particle accelerators are often thin metal foils deposited on a backing material like aluminium. In reactor irradiations, thicker targets are generally used to achieve higher yields. Liquid or gaseous targets can be held in polypropylene containers which have a low particle capture cross-section. Often targets have to be enriched with the particular isotope being transformed; this minimizes formation of other radionuclides which would reduce the specific activity of the desired species.

The formation of radionuclides is associated with a reaction energy, Q, which is a consequence of the conservation of energy. For a process X(a,b)Y, the reaction energy will be

$$Q = (M_X + M_a - M_Y + M_b)\, c^2 \tag{6.4}$$

When Q is positive a portion of the nuclear mass is converted to the kinetic energy of the products, e.g., $^2\mathrm{H}(n,\gamma)^3\mathrm{H}$ $Q = 6.26\,\mathrm{MeV}$. A negative value, e.g., $^{19}\mathrm{F}(p,\alpha)^{16}\mathrm{O}$ $Q = -8.12\,\mathrm{MeV}$, implies that there is a minimum threshold energy required for the reaction to occur. This can be calculated as

$$E_T = -Q\left(1 + \frac{M_a}{M_X}\right) \tag{6.5}$$

6.3.1 Formation of radionuclides in reactors

Key Point: Neutrons are directed at a target; the rate of conversion is a function of neutron capture rate and the decay rate of the product radionuclides.

The formation of radionuclides in nuclear reactors involves bombarding the target with a flow of neutrons. The neutrons are initially produced by so-called neutron generators which operate by directing a beam of deuterons at a solid target containing tritium, $^3\mathrm{H}(d,n)^4\mathrm{He}$. The neutrons produced have high kinetic energy, *ca.* $14\,\mathrm{MeV}$, and are known as ***fast neutrons***. If fast neutrons were directed at a target nucleus, they would have sufficient energy to eject two neutrons, creating a neutron-deficient radionuclide. However, the flow of fast neutrons can be ***moderated*** to reduce their kinetic energy by passing them through a material like boron, graphite, or water. The kinetic energy of the neutrons is reduced until they are in thermal equilibrium with the moderator; the relationship between their kinetic energy and temperature is given by the relationship:

$$E = \frac{3}{2}k_\mathrm{B}T \approx 4 \times 10^{-21}\,\mathrm{J} \tag{6.6}$$

where $k_B = 1.38 \times 10^{-23}\,\mathrm{J \cdot K^{-1}}$ (the Boltzmann constant). These neutrons, known as ***thermal neutrons***, bring about (n, γ) and (n,p) reactions which produce neutron-rich nuclei.

If we think about the production of a radionuclide, the general case could be represented as

$$Y \xrightarrow[\sigma]{\text{n}} N \xrightarrow{\beta-} N'$$

In this scheme, incident neutrons interact with the target, Y, by an amount proportional to the reaction cross-section, forming a product nuclide, N. However, this same nuclide decays to a daughter product, N', at a rate governed by its decay constant, λ. This is analogous to a secular equilibrium (half-life of the parent is much longer than the daughter) and we see that an identical kinetic treatment can be applied to determine the yield of a nuclear reaction.

First, we define the rate of formation of N in terms of the particle flux, reaction cross-section and number of target nuclei:

$$\frac{dN}{dt} = \phi\sigma n \tag{6.7}$$

We know from Chapter 5 that the rate of decay is given by λN so the net rate of production will be

$$\frac{dN}{dt} = \phi\sigma n - \lambda N \tag{6.8}$$

Integrating this first-order differential equation between appropriate limits and multiplying through by λ gives the activity of the radionuclide:

$$A \equiv \lambda N = \phi\sigma n(1 - e^{-\lambda t_{\text{irr}}}) \tag{6.9}$$

The activity obtained from Eq. (6.9) is really specific activity as it will have units of becquerel per cm^3. The time t_{irr} is known as the ***irradiation time*** and is taken as the duration of the neutron flux.

On completion of the irradiation a period of time passes before the activity can be measured; this is known as the ***cooling time***, t_{cool}, during which a portion of the radionuclide will have decayed:

$$A = \phi\sigma n(1 - e^{-\lambda t_{\text{irr}}})e^{-\lambda t_{\text{cool}}} \tag{6.10}$$

In cases where the half-life of the product is long, $\lambda t_{\text{irr}} \ll 1$ and the term $(1 - e^{-\lambda t_{\text{irr}}}) \approx \lambda t_{\text{irr}}$. In this case, Eq. (6.9) reduces to $N = \varphi\sigma n t_{\text{irr}}$. Conversely, if the product atoms have a very short half-live, $\lambda t_{\text{irr}} \gg 1$ and the term $(1 - e^{-\lambda t_{\text{irr}}}) \approx 1$ which means Eq. (6.9) becomes $A = \varphi\sigma n$. For most practical purposes, the target is irradiated for a length of time at least equal to a few half-lives of the product. Further irradiation beyond this time means the term in brackets in Eq. (6.9), known as the **saturation factor**, approaches unity.

A simple example of this is yttrium-90, a common beta emitter used in radiotherapy. Although it can be prepared form strontium-90 (obtained from uranium), it can also be formed by neutron bombardment of naturally occurring yttrium-89:

$$^{89}_{39}Y \xrightarrow[\sigma=1.3b]{n} \, ^{90}_{39}Y \xrightarrow[\lambda=0.011\,h^{-1}]{\beta^-} \, ^{90}_{40}Zr$$

To obtain a specific activity in MBq/mg, we calculate the number of target nuclei in 1 mg of pure yttrium-89:

$$n = \frac{0.001 \times 6.02 \times 10^{23}}{89} = 6.76 \times 10^{18} \text{ nuclei/mg}$$

If the neutron flux is $1.0 \times 10^{14} \text{ cm}^{-2}\text{ s}^{-1}$, the reaction cross-section is 1.3 b and the decay constant is $0.26\,\text{day}^{-1}$, the activity would be

$$A = (1 \times 10^{14}) \times (1.3 \times 10^{-24}) \times (6.74 \times 10^{18}) \times (1 - e^{-0.26 \times 10})$$
$$= 814 \text{MBq/mg}$$

If we plot the specific activity as a function of irradiation time (Figure 6.2), we see that this value is close to its saturation value.

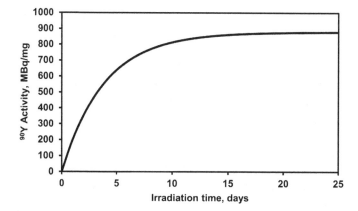

Figure 6.2 Formation of yttrium-90. As irradiation time increases, the activity of the product begins to level off to a saturation value.

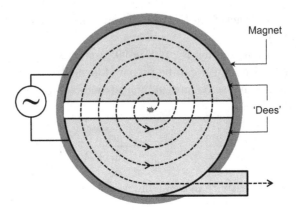

Figure 6.3 A cyclotron. Energetic particles are accelerated through a circular path and directed toward a target by the beamline.

6.3.2 Formation of radionuclides in cyclotrons

Key Point: Charged, energetic particles are directed at a thin, metallic target. Neutron-deficient (positron emitting) radionuclides are produced.

Radioisotopes can be produced using two types of particle accelerators: ***linear accelerators*** (linacs) and ***circular accelerators*** (cyclotrons). Both types of instrument have a source of charged particles, usually protons, deuterons, or alpha particles, which are accelerated to very high energies using a high frequency alternating voltage. The stream of energetic particles is then directed at a target material, bringing about the nuclear transformation. In a cyclotron, the particle stream originates at the center of a large circular vacuum chamber which contains two D-shaped metal electrodes sandwiched between two static electromagnets. As the particles are generated, they are accelerated in a circular path to velocities approximately three-quarters of the speed of light (Figure 6.3). They exit the cyclotron *via* a beamline which directs the particles to the target.

In terms of the kinetics of the process, the equations developed in above section can be equally applied to the formation of radionuclides in cyclotrons. The only modification of Eq. (6.9) *et seq.* which is necessary is the term for incident particle flux. If a target of thickness x is bombarded by a beam of particles of charge z at a current of I A, the particle flux will be

$$\phi = \frac{6.24 \times 10^{18} I}{z} \qquad (6.11)$$

For example, the bombardment of magnesium-26 with deuterons brings about the reaction:

$$\ce{^{26}_{12}Mg} \xrightarrow[\sigma=25\text{mb}]{d^+} \ce{^{24}_{11}Na} \xrightarrow[\lambda=0.047\,h^{-1}]{\beta^-} \ce{^{23}_{11}Na}$$

If the magnesium foil contains 11.0% magnesium-26 and the density is 1.74g/cm^3, the number density will be

$$n = \frac{1.74 \times 6.02 \times 10^{23} \times 0.11}{24.3} = 4.74 \times 10^{21}\text{nuclei/cm}^3$$

If the magnesium is irradiated by a beam of 22 MeV deuterons (which have $\sigma = 25\,\text{mb}$) for 2 h at 1×10^{-4}A the particle flux will be

$$\phi = \frac{6.24 \times 10^{18} \times 1 \times 10^{-4}}{1} = 6.24 \times 10^{14}$$

The activity will be given by Eq. (6.9), remembering to include the thickness of the foil, x:

$$A = (6.24 \times 10^{14}) \times (25 \times 10^{-27}) \times (4.74 \times 10^{21}) \times 0.01 \times (1 - e^{-0.047 \times 2})$$

$$= 6.64 \times 10^7 \text{Bq}$$

In most respects, production of radioisotopes by cyclotrons is preferred over nuclear reactor facilities. Firstly, there is much less nuclear waste produced by cyclotrons, and that which is produced is less hazardous than waste from comparable processes in a nuclear reactor. Secondly, the reactions in cyclotrons are easier to control — there is no risk of an uncontrolled nuclear reaction. Thirdly, cyclotrons can be installed at the point of use (e.g., in a hospital), eliminating the need for transport of radioactive material between sites.

6.4 Models of Nuclear Reactions

Key Point: There are three models used to explain nuclear reactions: the optical model, the liquid-drop model, and the direct interaction model. These are all largely empirical models based on observed phenomena.

Although our understanding of the nucleus has greatly improved since Rutherford's pioneering work, there is still no unifying theory of nuclear structure. So, in order to discuss nuclear reactions, we employ what could be described as "predictive models," that is, models which appear to fit certain experimental observations.

The first model is the ***optical model***. When light is incident on a crystal ball, it is scattered in various directions without any loss of energy. If light is directed at a black crystal ball, it is absorbed with no scattering. A crystal ball between these two extremes will reflect a portion of the light and absorb a portion. By analogy, when incident particles are directed at a target nucleus, a portion of them is absorbed and a portion is scattered. In the optical model, the physics of optics is used to develop an approach to understand the likelihood of a successful nuclear collision. Its mathematical formalism is based upon the potential well model of the nucleus, which contains terms for the absorption and scattering of particles. Although this model closely follows experimental results, it is purely superficial — it does not explain the events which occur during the nuclear reaction such as how the particle transformations are brought about.

In the ***liquid-drop model*** of nuclear reactions we employ a liquid-drop model of nuclear structure discussed in Chapter 3. In this approach, we propose that when an incident particle bombards the target nucleus, the kinetic energy of the former, when combined with the binding energy of the nucleus, creates excitation energy. The excitation energy is shared across the entire nucleus, with all nucleons moving with increased kinetic energy. At some point, a nucleon will have sufficient kinetic energy to overcome the binding energy and it escapes the nucleus. With the departure of a nucleon, there is now more excitation shared across the remaining nucleons and the process continues, with further expulsion of nucleons until the residual excitation energy is less than the binding energy of a nucleon. Any excess excitation energy remaining at this stage is released as gamma ray photons. The model makes use of the idea of a ***compound nucleus*** — an intermediate formed when the energy of the incoming particle is shared across the target nucleus. For example, tungsten-184 undergoes alpha particle capture to form osmium-186 and two neutrons *via* a tungsten-188 compound nucleus:

$$^{184}_{74}\text{W} + {}^{4}_{2}\text{He}^{2+} \rightarrow \left[{}^{188}_{76}\text{W}\right]^{*} \rightarrow {}^{186}_{76}\text{Os} + 2{}^{1}_{0}n$$

This nuclear reaction illustrates an important feature of the liquid-drop model. In general, neutrons tend to be expelled rather than protons, mainly because there are usually a larger number of neutrons in a nucleus, and also because neutrons can more readily cross the potential barrier than protons.

Finally, in the ***direct interaction model***, we suppose that the incident particle collides with only a few nucleons in the target nucleus, some of which may be directly ejected. Thus, the kinetic energy of the incident particle is not shared over the entire nucleus, but rather with just a few nucleons. It is also possible that the incident particle only loses a portion of its energy on collision with the nucleus and emerges after the initial collision. This model does not support the involvement of a compound nucleus; instead it proposes the involvement of *stripping or pickup* processes. In striping processes, a portion of the incident particle interacts with the nucleus, stripping some of its mass/energy, leaving a smaller particle to emerge. The most common example of this is the *Oppenheimer–Phillips process* (deuteron stripping reaction) in which a deuteron is incident upon a nucleus, increasing its mass by one and reemerging as a proton, e.g.,

$$^{12}_{6}\text{C} + \text{d} \rightarrow \, ^{13}_{6}\text{C} + \text{p}$$

In the pickup process, the projectile particle gains nucleons from the nucleus, e.g.,

$$^{20}_{10}\text{Ne} + \text{d} \rightarrow \, ^{19}_{9}\text{F} + \, ^{3}_{2}\text{He}$$

The nucleus of the particle produced, ^{3}He, is known as a *helion*, and the complete atom is one of two stable isotopes of helium. Helium-3 has a relatively low terrestrial abundance but has nevertheless been proposed as a fuel for second-generation fusion power plants.

6.5 Nuclear Fission

Key Point: Nuclear fission involves the splitting of a large, heavy nucleus into roughly equal nuclei with the release of large amounts of energy.

Nuclear fission involves the splitting of a heavy nucleus into roughly equal parts with the concomitant release of large amounts of energy (*ca.* 200 MeV per fission[b]). A characteristic of these reactions is that the total mass of the products is always less than that of the reactants.

Any nuclide which is capable of undergoing fission after capture of a high-energy neutron is described as being *fissionable*, e.g., uranium-235:

$$^{235}_{92}\text{U} + \text{n} \rightarrow \, ^{140}_{56}\text{Ba} + \, ^{93}_{36}\text{Kr} + 3\text{n}$$

In this case, the three neutrons produced by the fission reaction can induce fission in other uranium-235 nuclei, forming a *nuclear chain reaction*. The neutrons released at the instant of fission are known as *prompt*

---

[b]Roughly equivalent to 80 TJ/kg, 20 kilotons of TNT or the energy released from combustion of 14000 kg of coal.

neutrons and the average number of prompt neutrons is characteristic of a particular fission reaction. In addition to these prompt neutrons, *delayed neutrons* may be released following the beta decay of daughter nuclei (e.g., the decay of rubidium-93 to strontium-92 involves release of a neutron).

The products of the fission of uranium-235 need not always be barium-140 and krypton-93; careful investigation has shown that the mass of the fragments can span from isotopes of zinc to gadolinium, with maxima at around $A = 95$ and 135. This pattern of fragmentation is referred to as *binary fission*; a further process, *ternary fission*, also occurs at a much lower incidence and typically produces helium-4 and tritium.

The mechanism for nuclear fission is satisfactorily described by the liquid-drop model. Initially, when uranium-235 captures a neutron, a compound nucleus is formed which then breaks apart to give the fission products and *ca.* three neutrons:

$$^{235}_{92}\text{U} + \text{n} \xrightarrow{\sigma = 98.3\,\text{b}} {}^{235}_{92}\text{U}^* \xrightarrow{10^{-12}\text{s}} \text{X} + \text{Y} + \text{n}$$

When a fissile nucleus captures a neutron, the resultant excitation energy is sufficient to deform the compound nucleus into a double-lobed structure; the nuclear force weakens to the extent where it can no longer support the compound nucleus and fragmentation occurs at this stress point (Figure 6.4). We can provide a rough mathematical rationale for this mechanism by calculating the potential energy of the compound nucleus when the nuclear force falls to zero.

Taking rubidium-93 and caesium-143 as the fission products, we first calculate their nuclear radii (*c.f.* Eq. (3.9)):

$$r(\text{Rb}) = (1.4 \times 10^{-15}) \times (93)^{1/3} = 6.3 \times 10^{-15}\text{m}$$
$$r(\text{Cs}) = (1.4 \times 10^{-15}) \times (143)^{1/3} = 7.3 \times 10^{-15}\text{m}$$

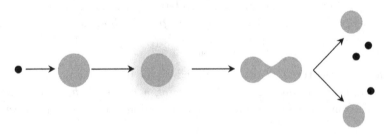

Figure 6.4 Nuclear fission. The liquid-drop model predicts the formation of a compound nucleus which becomes deformed and splits into fission products and neutrons.

Table 6.1 Energy Release for Fission of Uranium-235

Process/Stage	Energy, MeV
Prompt energy relase	
Kinetic energy of fission products	164.6 ± 4.5
Kinetic energy of neutrons	4.9 ± 0.5
Gamma decay energy	7.0 ± 0.5
Total	176.5 ± 5.5
Delayed energy release	
Kinetic energy of beta particles	6.5 ± 1.5
Neutrino release	10.5 ± 2.0
Gamma decay energy	6.5 ± 1.5
Total	23.5 ± 5.0
Overall total	200 ± 6.0

The distance between the two nuclei in the compound nucleus will be the sum of these values, 13.6×10^{15} m. Employing Coulomb's law we can estimate the potential energy:

$$U = k\frac{q_1 q_2}{r} = \left(1.44\text{eV} \cdot \text{nm}^{-1}\right) \times \frac{37 \times 55}{1.36 \times 10^{-5}\text{nm}} = 212\,\text{MeV}$$

This value is in good agreement with the binding energy for a uranium-235 nucleus (200 MeV), which would seem to imply that our model is at least a good approximation.

The balance sheet in terms of energy liberated by nuclear fission is summarized in Table 6.1. The amount of initial activation energy is in the region of 6.2 MeV; when this is satisfied, the compound nucleus is formed. Neglecting the kinetic energy of the neutron (which is relatively very small), the energy of the compound nucleus is about 6.5 MeV. Thus, the energy requirement is exceeded by the energy obtained by adding a neutron to uranium-235. This is consistent with the large slow neutron cross-section for uranium-235 (*ca.* 585b). Using Eq. (6.4) we can evaluate Q for the fission reaction as 165 MeV. This combined with the energies associated with prompt neutron release, prompt gamma rays, and radioactive decay account for the remaining 35 MeV.

6.6 Nuclear Fusion

Key Point: Nuclear fusion involves the joining of two high-energy nuclei to form a more stable product. This forms the basis of stellar nucleosynthesis.

Nuclear fusion is in many respects the opposite of nuclear fission; a fusion reaction is characterized by the formation of a new nucleus by colliding high-energy nuclei. While fission can occur naturally on Earth, fusion cannot; it does, however, form the basis of the energy-generating process in main-sequence stars, such as the Earth's Sun. In order for fusion to occur, the repulsive Coulomb barrier must be overcome, which necessitates an input of energy. One approach to meet this energy demand is to increase the thermal energy of the nuclei either artificially, as occurs in laser-induced fusion for example, or naturally through the heat released by nuclear reactions. The use of thermal energy to overcome the Coulomb barrier originated with the work of G. Gamow, whose theory provides the probability of two nuclei becoming sufficiently close for the nuclear force to overcome the Coulomb repulsion.

First, we consider the most elementary fusion reaction which occurs in our Sun, the fusion to two hydrogen nuclei:

$$\mathrm{^1_1H + {}^1_1H \rightarrow {}^2_1H} + e^+ + \nu_e$$

We can calculate the size of the potential energy barrier by recognizing that in order for the Coulomb repulsive force to be overcome the two nuclei must be within 1.0×10^{-14} m (the range of the nuclear force). Taking the charge of the hydrogen nuclei as $q_1 = q_2 = e = 1.6 \times 10^{-19}$ C and Coulomb's constant as 8.99×10^9 Nm·C^{-2}, we obtain

$$U = k\frac{q_1 q_2}{r} = 8.99 \times 10^9 \times \frac{\left(1.6 \times 10^{-19}\right)^2}{1.0 \times 10^{-14}} = 2.3 \times 10^{-14} \mathrm{J}$$

If we halve this value for a single hydrogen nucleus and use Eq. (6.6), we can estimate the temperature required for the fusion reaction:

$$T = \frac{2E}{3k_{\mathrm{B}}} = \frac{2 \times 1.1 \times 10^{-14}}{3 \times 1.38 \times 10^{-23}} = 5.3 \times 10^8 \mathrm{K}$$

If we compare this to the temperature of the Sun's core, *ca.* 1.6×10^7 K, we see that our estimate is higher than that possible in the Sun! However, we need to recognize that this calculation is based on the classical Maxwell–Boltzmann distribution which does not take into account quantum mechanical effects. If we treat hydrogen nuclei (protons) as waves, then quantum mechanical tunneling makes their passage to outside the potential barrier possible. After that, the fusion reaction is feasible at the temperature of the Sun's core.[c]

The cross-section of the proton–proton reaction is very low, around 10^{-33}b, suggesting that this reaction is the rate-limiting step in the overall fusion process occurring in the Sun. As only a small number of deuterons are formed, the most likely second step is the reaction of a deuteron with a hydrogen nucleus:

$$\ce{^1_1H} + \ce{^2_1H} \rightarrow \ce{^3_1He} + \gamma$$

Finally, two helium-3 nuclei fuse to form helium-4 which is the final step in the fusion reaction:

$$\ce{^3_2He} + \ce{^3_2He} \rightarrow \ce{^4_2He} + 2\ce{^1_1H} + \gamma$$

The net conversion of four protons to helium is known as the ***proton–proton cycle*** and is characteristic of stars which are of approximately the same size and composition as our Sun.

In larger stars, where the temperature and pressure are higher, an alternative fusion process, known as the ***CNO cycle*** (carbon–nitrogen–oxygen cycle) occurs. The fusion of carbon can occur at temperatures $\geq 8 \times 10^8$ K and can be summarized as a series of six fusion processes:

$$^{12}_{6}\text{C} + {}^1_1\text{H} \rightarrow {}^{13}_{7}\text{N} + \gamma$$
$$^{13}_{7}\text{N} \rightarrow {}^{13}_{6}\text{C} + e^+ + \nu_e$$
$$^{13}_{6}\text{C} + {}^1_1\text{H} \rightarrow {}^{14}_{7}\text{N} + \gamma$$
$$^{14}_{7}\text{N} + {}^1_1\text{H} \rightarrow {}^{15}_{8}\text{O} + \gamma$$
$$^{15}_{8}\text{O} \rightarrow {}^{15}_{7}\text{N} + e^+ + \nu_e$$
$$^{15}_{7}\text{N} + {}^1_1\text{H} \rightarrow {}^{12}_{6}\text{C} + {}^4_2\text{He}$$

Note that in this series of reactions, carbon-12 is initially consumed and then replenished, suggesting that its role is catalytic. Although the CNO cycle only accounts for approximately 0.8% of the Sun's fusion output, as it progresses through its life cycle and the core temperature increases, the activity of the CNO cycle will increase.

---

[c]The Sun's incredibly high temperature comes from the fact that when matter condenses, gravitational energy is released which is converted to kinetic energy, which in turn increases temperature and pressure. At a critical density (known as the Chandrasekhar limit), the temperature is high enough to promote ionization of hydrogen atoms, releasing protons and electrons.

The CNO cycle, first proposed by Carl von Weizsäcker in 1839, lead to the recognition of a process now referred to as *stellar nucleosynthesis*, the process through which lighter nuclei are converted into heaver products. Stellar nucleosynthesis is responsible for the formation of elements up to iron, which is at the peak of the binding energy curve. Beyond iron, fusion ceases to release energy. For this reason, stars which are progressing toward the end of their life cycle accumulate an iron core, which is expelled on collapse of the star. Heaver isotopes are formed through slow neutron capture (the s-process), while very heavy isotopes are thought to be formed through rapid neutron capture (the r-process).

Chapter Summary

- Nuclear reactions bring about the transmutation of one isotope to another, usually with the release of energy and radiation.
- The likelihood of a nuclear reaction occurring can be described by the reaction cross-section which has units of barn.
- Artificial radionuclides can be formed in nuclear reactors or accelerators such as cyclotrons. Each approach produces commercially valuable radionuclides for research, healthcare, and technological use.
- There are three main models describing nuclear reactions: the optical model, the liquid-drop model, and the direct interaction model. These are predictive models which are in keeping with experimental observations.
- Nuclear fission involves the splitting of heavy nuclei by fast, thermal neutrons. The process produces smaller nuclei and large amounts of energy. Fission reactions form the basis of the nuclear power industry.
- Nuclear fusion involves the formation of a new nucleus by combining smaller nuclei. This process requires considerable thermal energy to overcome the repulsive Coulomb barrier and is the main form of reaction occurring in main sequence stars.

Review Questions

(1) Suggest the likely identity of the product nuclide of the following nuclear reactions: (a) ^{20}Ne(α,γ); (b) ^{15}N(p,α); (c) ^{10}B(n,α); (d) ^{23}Na(n,β^-).

(2) The nuclear transformation ^{14}N$(\alpha,p)^{17}$O was first reported by Rutherford in 1917. Calculate the energy change for this reaction given the following atomic masses: nitrogen, 14.00307; helium, 4.00260; hydrogen, 1.007825; and oxygen, 16.99913.

(3) Seaborgium-263 is produced by the bombardment of californium-249 with oxygen-18 nuclei. Provide a balanced nuclear equation for this process and suggest the identity of the daughter nuclide following alpha decay of the seaborgium-263.

(4) Pyrex glass contains significant quantities of boron-10 which has a high cross-section for neutron capture. Suggest why glass vessels are not used to hold target material in a nuclear reactor.

(5) Calculate the irradiation time required to produce 0.06 g cobalt-60 ($t_{1/2} = 5.27$ years) with an activity of 37 MBq by exposing cobalt-59 to a neutron flux $5 \times 10^{13} \text{cm}^{-2}\,\text{s}^{-1}$ with a cross-section of 37 b.

(6) A manganese foil, $10\,\text{mg/cm}^2$, is irradiated with alpha particles in a cyclotron for 1 h operating at $1\,\mu\text{A}$. If the cross-section for the $^{55}\text{Mn}(\alpha,2n)^{57}\text{Co}$ reaction is 200mb, what will be the specific activity of the cobalt-57 ($t_{1/2} = 270$ days)?

(7) The reaction $^{65}\text{Cu}(n,2n)^{64}\text{Cu}$ can be mediated by short exposure to thermal neutrons at $1 \times 10^7 \text{cm}^{-2}\,\text{s}^{-1}$. If 2.0×10^5 atoms of copper-64 were produced from copper foil ($20\,\text{mg/cm}^2$), estimate the reaction cross-section for the process.

(8) Two deuterons can fuse to form an alpha particle. Using Coulomb's law, estimate the kinetic energy and hence the temperature required for the fusion process. Take $k = 9.0 \times 10^9 \text{Jm·C}^{-2}$ $q = 1.6 \times 10^{-19}\text{C}$ and $r = 2 \times 10^{-15}\text{m}$.

(9) Calculate the overall value of Q for the Sun's proton– proton cycle (i.e., $^1\text{H} + ^1\text{H} \rightarrow {}^2\text{H}$; $^1\text{H} + ^2\text{H} \rightarrow {}^3\text{He}$ and $^3\text{He} + ^3\text{He} \rightarrow {}^4\text{He} + 2^1\text{H}$) taking the following atomic masses: $^1\text{H} = 1.007825$; $^4\text{He} = 4.002602$.

(10) Find the fission energy for the reaction

$$^{235}_{92}\text{U} + \text{n} \rightarrow {}^{141}_{56}\text{Ba} + {}^{92}_{36}\text{Kr} + 3\text{n}$$

taking the following masses: $\text{n} = 1.008665$; $^{235}\text{U} = 235.043915$; $^{141}\text{Ba} = 140.9139$; $^{92}\text{Kr} = 91.8973$.

Chapter 7

Radioactivity at Work

"I do not believe in the commercial possibility of induced radioactivity."

John Haldane

Radioactive sources are used in a variety of environments, ranging from metalworking to delicate instruments on orbiting satellites. Those engaged in handling radioactive substances are required to have varying levels of awareness, depending on the nature of their work. In this chapter, we discuss the main practical issues surrounding working with radioactivity. On completion of this chapter and the associated questions, you should:

- Be able identify external and internal sources of radiation, and be able to carry out simple dose calculations.
- Have an overview of the main legislation relating to use of radioactive substances.
- Understand the principles of monitoring radiation, both qualitatively and quantitatively.

7.1 Why Use Radioisotopes?

Key Point: Radioactive sources used in industry, education, or research are classified as sealed or unsealed.

In the laboratory, radioisotopes find application in a variety of fields, but it is often a legal requirement that the use of a radioisotope must have clear advantages over other methods. In some instances, a radioisotope must be used — for example, when investigating the inorganic chemistry of uranium. In other cases, particularly when labeling compounds or macromolecules, the use of non-radioactive isotopes (e.g., ^{2}H), or even fluorescent labels may be preferable. The use of non-radioactive isotopes is common in analytical chemistry — internal standards in GC/MS or D_2O in ^{1}H-HMR studies. In these cases, the only legislative requirements are for good laboratory practice and safe disposal of chemical waste. For radioactive isotopes, specific legislation must be followed, which becomes increasingly stringent depending on the activity of the radionuclide.

Radioactive materials used in laboratories or in industry can fall into one of two categories: sealed and unsealed sources. **Sealed sources** are common in teaching laboratories, or as part of an instrument or plant machinery. The definition of a sealed source varies slightly between legislative regions, but in general, it can be regarded as a radioactive material (<400 kBq) encapsulated within, or bonded to, a material, which is strong enough to resist damage due to normal use. It should be mounted in a tamper-resistant holder and the entire unit should conform to international standards.[a] **Unsealed sources** are any solid or liquid radioactive material that can be purchased or synthesized. These are more common in research laboratories and carry more stringent safety requirements than sealed sources.

7.2 External Exposure to Radiation

Key Point: External radiation has different effects on various tissues; this is represented by the idea of an effective dose.

All forms of radiation emitted from a point source follow an inverse square law, such that the intensity of the radiation decreases with the square of the distance:

$$I \propto \frac{1}{d^2} \qquad (7.1)$$

The inverse square law would apply to a sample of radioactive material sitting on the bench, for example, in that it will emit radiation equally in all directions. If the radioactive source is shielded to produce a narrow beam (collimation), then the beam of radiation will travel in a roughly linear path.

---

[a]ISO2919:1999 — see http://www.iso.org/iso/iso_catalogue/catalogue_ics.

In the SI system, exposure to either X-rays or gamma radiation is given in coulombs per kilogram and is referred to as the **X-unit** (exposure dose). This is the quantity of radiation that produces ions carrying a charge of one coulomb per kilogram of dry air at standard temperature and pressure. When the radiation encounters matter, a portion of the radiation will be absorbed, with the extent of absorption varying between different materials. The SI unit of **absorbed dose**, D, is the gray (Gy), which is the absorption of one joule of energy per kilogram of mass.[b] However, for biological absorption the concept of the **equivalent dose**, H, was introduced to take into account absorption by different tissues. The two quantities are related by the radiation weighting factor, w_R, a constant specific to the type of radiation:

$$H = D \cdot w_R \tag{7.2}$$

For beta particles, gamma rays and X-rays, $w_R = 1$, and for alpha particles and neutrons, $w_R = 20$. If there is exposure to more than one type of radiation, the equivalent dose is calculated as the function:

$$H = \sum_R D \cdot w_R \tag{7.3}$$

The unit of the equivalent dose is the sievert (Sv), although other non-SI units are also in common use, particularly in the USA (see Table 7.1). The **effective dose**, E, takes into account the different sensitivities different organs have to radiation. It is obtained by summation of the equivalent

Table 7.1 Units in Radiological Protection

Measurement	SI Unit	Non-SI Unit	Conversion
Activity	Bq	Ci	$1 \text{ Ci} = 3.7 \times 10^{10} \text{ Bq}$
Exposure	X-unit	R	$1 \text{ X} = 3876 \text{ R}$
Absorbed dose	Gy	rad	$1 \text{ Gy} = 100 \text{ rad}$
Equivalent dose	Sv	rem	$1 \text{ Sv} = 100 \text{ rem}$
Effective dose	Sv	rem	$1 \text{ Sv} = 100 \text{ rem}$
Dose rate	Gy h^{-1}	rad h^{-1}	$1 \text{ Gy/h} = 100 \text{ rad/h}$

R, Röentgen, rad, radiation-absorbed dose; rem, Röentgen equivalent man.

---

[b]Radiation exposure and absorbed dose are related since exposure is based on absorption by air. However, as living tissue is much denser than air, it will absorb more energy per unit mass. Consequently, 1 X-unit $\approx$ 34 Gy.

Table 7.2 Tissue-Weighting Factors
for Calculation of Effective Dose

Tissue	W_T
Gonads	0.25
Breast	0.15
Red bone marrow	0.12
Lung	0.12
Thyroid	0.03
Bone surfaces	0.03
Remainder	0.30

doses multiplied by tissue-weighting factors (Table 7.2):

$$E = \sum_T H \cdot w_T \qquad (7.4)$$

Finally, we consider that **dose rate**, $\dot{D}$, which takes into account the duration of the exposure to radiation. External exposure to alpha and beta particles is less likely to cause serious damage as they have low penetrating power. Some beta emitters ("hard beta emitters"), such as phosphorous-32, may pose a greater risk to uses as these have higher energies. The dose rate for such an emitter ten centimeters away from the user can be calculated from Eq. (7.5), where A is the activity (in MBq) of the source:

$$\dot{D} = 760A \qquad (7.5)$$

Gamma rays (and X-rays) pose a much more significant risk and the dose rate is evaluated using **specific dose rate constants**, Γ (Table 7.3):

$$\dot{D} = \frac{\Gamma A}{d^2} \qquad (7.6)$$

where d is the distance in meters (see calculations, which follow).

External exposure to radiation can be minimized by placing a physical barrier between the source and the handler. This is usually referred to as **shielding** and a variety of materials can be employed, each with their own specific characteristics, which make them appropriate for a particular type of radiation. Exposure to alpha and beta radiation does not represent a serious external radiological hazard, except for hard beta emitters. In these cases, a Perspex shield is usually employed. With gamma radiation, however, external exposure can be significant and a more comprehensive consideration of shielding is required. The attenuation of gamma rays is considered in terms of **half-value layer** (HVL), which is the thickness of a

Table 7.3 Specify γ-Decay Constants for Selected Radionuclides

	X-m^2 (h·MBq)$^{-1}$	Γ Gy·m^2 (h·MBq)$^{-1}$	R·m^2 (h·Ci)$^{-1}$
^{137}Cs	2.30×10^{-9}	7.82×10^{-8}	0.330
^{51}Cr	1.11×10^{-10}	3.77×10^{-9}	0.016
^{60}Co	9.19×10^{-9}	3.2×10^{-7}	1.321
^{125}I	4.87×10^{-10}	1.66×10^{-8}	0.070
^{131}I	1.53×10^{-9}	5.20×10^{-8}	0.220
^{22}Na	8.36×10^{-9}	2.84×10^{-7}	1.200
^{24}Na	1.28×10^{-8}	4.35×10^{-7}	1.840
^{65}Zn	1.88×10^{-9}	6.39×10^{-8}	0.270

specified absorber that reduces the exposure rate by half. Different materials have different HVLs and these can be used to calculate the thickness of shielding required to reduce exposure rate to a safe level.

A ballpark estimate of the depth of shielding required can be easily calculated as follows. Suppose, we had a cobalt-60 gamma ray source with an activity of 18.5 GBq held 45 cm away from the user. Using Eq. (7.6), we can determine the unshielded dose rate:

$$\dot{D}_0 = \frac{\left(3.12 \times 10^{-7} \frac{\text{Gy·m}^2}{\text{hr·MBq}}\right) \times 18500\,\text{MBq}}{(0.45^2\,\text{m})^2} = 2.85 \times 10^{-2}\,\text{Gy/h}$$

If we take a safe dose of gamma radiation as 9.6×10^{-5} Gy/h, the required attenuation will be:

$$\frac{9.60 \times 10^{-5}\,\text{Gy/h}}{2.85 \times 10^{-2}\,\text{Gy/h}} = 0.0034$$

From the definition of HVL, we can surmise that n HVLs will reduce the radiation dose by a factor of $1/2^n$, so therefore:

$$\frac{1}{2}^n = 0.0034 \quad \text{and} \quad n\log\frac{1}{2} = \log 0.0034 \therefore n = 8.2$$

We then multiply this result by the attenuation coefficient of the shielding material (lead $= 0.798$ cm^{-1} at 1000 keV[c]), which gives a thickness of 6.5 cm for lead.

---

[c]The attenuation coefficient is dependent on the energy of the gamma ray photon.

A more accurate calculation for a shielding material of thickness x is given by Eq. (7.7):

$$\dot{D} = \dot{D}_0 e^{-\left(\frac{(\ln 2)x}{\text{HVL}}\right)} \tag{7.7}$$

In which $\dot{D}$ is the radiation dose outside the shielding and $\dot{D}_0$ is given by Eq. (7.6). If we use this approach for the ^{60}Co problem, taking the HVL of lead as 1.2 cm, we can solve Eq. (7.7) by taking natural logarithms of both sides and rearranging for x:

$$\ln\left(9.6 \times 10^{-5}\frac{\text{Gy}}{\text{h}}\right) = \ln\left(2.85 \times 10^{-2}\frac{\text{Gy}}{\text{h}}\right) \times \left(-\frac{(\ln 2)x}{1.2\,\text{cm}}\right)$$

$$-\frac{\left[\ln\left(2.86 \times 10^{-9}\right) - \ln\left(2.85 \times 10^{-2}\right)\right] \times 1.2}{\ln 2} = x = 9.88\,\text{cm}$$

This value is higher than that obtained by the approximation method, but neither value should be taken at face value; any shielding is always assessed by a survey meter before being considered safe, and is continually monitored as part of the local rules (*c.f.* Section 7.4).

7.3 Internal Exposure to Radiation

Key Point: Internal dose of radiation depends on both the radioactive and biological half-life of the radionuclide.

Radionuclides can enter the body through inhalation, ingestion, or *via* the skin and their retention depends on the chemical nature of the material. The International Commission on Radiological Protection (ICRP) broadly states that the dose from internal exposure depends on four factors:

(1) Energy of emitted radiation: the dose is dependent on the amount of energy absorbed. For example, hydrogen-3 has decay energy of around 6 keV while that of iodine-131 is about 181 keV. The latter would potentially represent a greater risk.
(2) Type of radiation: Alpha and beta particles will be absorbed by the body if emitted internally, whereas gamma radiation will probably escape with relatively little absorption; therefore, for internal exposure, alpha, and beta particles are considered more hazardous.
(3) Distribution of the radionuclide: Some radionuclides concentrate in specific organ systems or tissues, increasing the dose received. For example, if Na^{125}I is ingested, the thyroid gland would be regarded as a *critical organ*.

(4) Rate of elimination: The biological half-life of the radioisotope is the rate, at which it is excreted from the body. Those substances with long biological half-lives, such as phosphorous-32, which accumulates in the bone, are regarded as particularly hazardous.

Due to the various factors outlined above, individual radioisotopes have an associated maximum permissible concentration in the environment, which reflects their perceived hazardousness. If a radioactive substance is released into the water supply, for example, there is a possibility it could be ingested by humans. The total amount of a given radionuclide in a person's body is referred to as the **body burden**, which is used to calculate doses from internal sources of radiation.

To illustrate the principle of body burden, consider an average 70 kg man who has consumed drinking water containing the maximum permissible concentration of tritium (111 Bq/mL).[d] Assuming that 60% of a person's body mass is water, this gives a potential reservoir of 42 kg for the tritium. First, to achieve unit agreement, we see that the units of mass must be converted to mL (1 g = 1 mL) and the body burden is therefore:

$$(42000\,\cancel{\text{mL}}) \times \left(111\frac{\text{Bq}}{\cancel{\text{mL}}}\right) = 4.66 \times 10^6 \text{Bq} \ (= 4.66\,\text{MBq})$$

As the body burden is 4.66×10^6 disintegrations per second, the energy absorbed per second is obtained by multiplying this result by the energy of the beta particle emitted by tritium (6 keV = 9.61×10^{-16} J):

$$\left(4.66 \times 10^6\,\frac{\cancel{\text{d}}}{\text{sec}}\right) \times \left(9.61 \times 10^{-16}\frac{\text{J}}{\cancel{\text{d}}}\right) = 4.48 \times 10^{-9}\text{J/sec}$$

Recalling that the dimensions of absorbed dose are joules per kilogram, we divide this result by the mass of the man to give the absorbed dose in Gy per second:

$$\frac{4.48 \times 10^{-9}\frac{\text{J}}{\text{sec}}}{70\,\text{kg}} = 6.40 \times 10^{-11}\,\text{J/sec} \times \text{kg} = 6.40 \times 10^{-11}\,\text{Gy/sec}$$

We can convert this result to the absorbed dose per year by multiplying by 3.16×10^7 sec/year, which would give an equivalent dose of 2.02 mSv/year

---

[d]It is a good idea to always include units as shown in these calculations as it often minimizes errors associated with unit conversion. It is essentially a form of dimensional analysis.

(*c.f.* Eq. (7.2) with $w_R = 1$ for β-particles), which is less than the maximum limit for occupational exposure (10 mSv/year). Of course, any internal dose of radiation should be treated seriously, particularly if there are toxicological effects in addition to those, which are due to radiation.

An obvious omission from this calculation was any account of the decrease in activity over time through radioactive decay and biological elimination. To adjust absorbed dose accordingly, we need to introduce steps, which describe radioactive decay and biological elimination. Conveniently, the latter generally obeys first-order kinetics and it can be shown that absorbed dose is given by

$$\dot{D} = \frac{\dot{D}_0}{\lambda_c} \left(1 - e^{-\lambda_c t}\right) \tag{7.8}$$

where λ_c is the sum of the decay constants (i.e., the radioactive decay constant and the rate constant for the elimination process). This is often summarized as the ***effective half-life,*** which is evaluated as

$$\frac{1}{t_{1/2e}} = \frac{1}{t_{1/2r}} + \frac{1}{t_{1/2b}} \quad \text{or} \quad t_{1/2e} = \frac{t_{1/2r} \times t_{1/2b}}{t_{1/2r} + t_{1/2b}} \tag{7.9}$$

where $t_{1/2e}$ is the effective half-life, $t_{1/2r}$ is the radioactive half-life, and $t_{1/2b}$ is the biological half-life. In the case of the tritium example used earlier, as it has a radioactive half-life of 4.5×10^3 days and a biological half-life of 12 days, the effective half-life would be

$$t_{1/2e} = \frac{(4.5 \times 10^3) \times 12}{(4.5 \times 10^3) + 12} = 11.96 \approx 12\,\text{days}$$

This is relatively short and therefore poses fairly minimal risk. Some other radionuclides, such as phosphorous-32, have biological half-lives measured in years, which obviously presents a much greater radiological hazard.

7.4 Biological Basis of Radiological Legislation

Key Point: The biological effects of radiation are used to define maximum exposure limits in different categories of radiation worker.

The biological effects of ionizing radiation are classified as being either ***stochastic*** or ***deterministic*** (non-stochastic or acute). Stochastic effects are those which arise by chance and usually present sometime after exposure to the radiation — for example, cancer or genetic abnormalities in offspring.

From a legislative point of view, it is very difficult to define thresholds of "safe levels" of radiation exposure with respect to stochastic effects — two people exposed to exactly the same dose of radiation could have completely different reactions; one could develop cancer for example, while the other may not. In the UK, the accepted maximum level of exposure for stochastic effects in a designated radiation worker is 20 mSv/year, while that for a member of the general public is 1 mSv/year. Protection against *in utero* effects in women of "reproductive capacity" is maintained by lowering the threshold to 13 mSv/year; if a pregnancy is confirmed, the maximum threshold is 1 mSv/year.

Deterministic effects are those which are apparently immediately on exposure to radiation and could include skin redness (erythema), hair loss, cataracts or death, depending on the dose, or radiation absorbed. These are much easier to legislate for as they can be based on empirical evidence. An example of this is the limit of radiation for the lens of the eye. The formation of a cataract is a deterministic effect with a threshold of 7.5 mSv. Taking the maximum exposure time for a radiation worker as 50 years, the limit for lens irradiation would be

$$\frac{7.5\,\text{mSv}}{50\,\text{yr}} = 150\,\text{mSv/year}$$

The limit for deterministic effects in other tissue types in radiation workers is 500 mSv/year. Therefore, provided that the limit for the irradiation of the lens is observed, the limit for other tissue types should never be reached.

7.5 Regulations in the United Kingdom

Key Point: Two key pieces of legislation are used in the UK to regulate the use of ionizing radiation and to minimize exposure. These are based on wider European legislation and are broadly similar to most other jurisdictions (e.g., the USA).

Within the UK, there are two key pieces of legislation — the *Ionising Radiations Regulation (1999)* (IRR99) and the *Radioactive Substances Act (1993)* (RSA93). The full documents are available to download from the Health & Safety Executive website. IRR99 provides a legal framework for minimizing the exposure of individuals to ionizing radiation. The guiding principle of IRR99 is the idea that exposure to ionizing radiation should be as low as reasonably possible

(ALARP). The main articles of these regulations are divided among seven sections:

(1) Provides the legal definitions for various terms, such as dose rate, etc.

(2) Outlines the general principles of radiation protection including risk assessment, restriction of exposure, dose limitation, and contingencies.

(3) Outlines the strategies for the management of radiation protection, such as the appointment of a radiation protection adviser (RPA).

(4) Presents the requirements for designated working areas, radiation protection supervisors (RPS), and establishing local rules.

(5) Addresses Designated Persons (those who work with radioactive sources) and the procedures for monitoring of exposure (dosimetry, medical monitoring, etc.).

(6) Links to the RSA93 and describes the regulations dealing with leak testing, record keeping, storage, and notification procedures for spillages/leaks.

(7) Outlines the duties and responsibilities of employees who use radioactive sources and various administrative details.

The regulations set dose limits for various categories of individual (Table 7.4), which consider the stochastic effects and deterministic effects of ionizing radiation.

The RSA93 focuses on the control of radioactive material and procedures for the disposal of radioactive waste. Before obtaining any radioactive material (open or sealed sources), it is necessary to obtain Certificates of Registration and Authorization from local authorities. This registration will specify the types of radionuclide permitted and the total activities allowed. As part of the registration process, evidence must be provided that

Table 7.4 Yearly Exposure Limits Defined in IRR99

Individual Classification	Exposure Limits (per year)	
Stochastic effects		
Radiation workers	20 mSv	2 rem
Members of the public	1 mSv	0.1 rem
Trainees	6 mSv	0.6 rem
Deterministic effects		
Radiation workers (lens of eye)	150 mSv	15 rem
Radiation workers (other tissues)	500 mSv	50 rem
Trainees (lens of eye)	50 mSv	50 rem
Trainees (other tissues)	150 mSv	15 rem
Women of reproductive age	13 mSv	1.3 rem
Pregnant women	1.3 mSv	0.13 rem

demonstrates safe storage of radioactive materials — for example, stored in a shielded, locked cabinet in a room with minimal footfall. The means of safe disposal of radioactive waste is also specified in the Certificates of Registration and Authorization. For example, aqueous waste can only be flushed down designed sinks for which the flow rate to the sewerage network is known, so that adequate dilution has been observed. Organic solvents, which contain radioactive materials, are usually incinerated in designated facilities. Solid radioactive waste with low activities can usually be disposed of at local waste dumps where they are immediately buried under other waste. If any waste has particularly high activities, their disposal must be overseen by a regulated disposal service.

7.6 Radiological Protection

Key Point: Legislation provides the main features of the local rules, which must be in place in every institution using ionizing radiation. These are enforced by a RPS.

Working with radioactive materials requires precautions beyond that of standard laboratory practice. These precautions must form part of the *local rules*, which describe how the work will be undertaken, what the nature of the hazards are, and the contingency plan should an accident occur. These local rules are derived from legislation and are generally written by a RPA.

The first step in radiological protection is the identification of designated working areas, which may be supervised or controlled, depending on the nature of the work. Since exposure to radiation could be from external and/or internal sources, the likelihood of exposure in a designated working area is considered from the point of view of the type of radionuclide being used. For example, if a gaseous compound labeled with iodine-125 is being prepared, the risk for internal exposure is much higher than if preparing a labeled solid or liquid compound. If there is the potential for exposure to an external dose at an instantaneous dose rate greater than 2 mSv/h, the working area becomes a *controlled area*. If the instantaneous dose rate is less than this threshold, the lab space would be considered a *supervised area*.

Regardless of its designation, any laboratory making routine use of radionuclides (a "hot lab") should be designed in such a way so as to minimize exposure to radiation. Lab benches should be covered in a material, which is easy to decontaminate (e.g., polyurethane or Perspex) and hard-to-clean corners should be avoided. Removable coverings (e.g., Benchkote®) have some advantages, especially if the lab is temporary, but their use greatly increases the volume of radioactive waste. Another alternative is to use large plastic trays lined with absorbent paper. An efficient fume cupboard is essential, as is a laboratory sink dedicated to the disposal of aqueous radioactive material. A shoe-change barrier at all doorways is also recommended.

7.7 Handling Radioactive Materials

Key Point: Radioactive materials, particularly unsealed sources, must be handled in designated areas following an approved procedure. A contingency plan for accidental spillage must form part of the local rules.

Radioactive materials are generally delivered in a sealed container, perhaps stored on dry ice depending on the nature of the substance. Certain containers may incorporate steel or lead shields, depending on the activity of the material. Radioactive solutions are usually packaged in borosilicate glass vials with a volume of seven milliliters and a screw cap with two seals. The inner seal is often a Teflon-faced rubber septum, which can be pierced by an appropriate gauge needle. If the material is dissolved in benzene or toluene, the inner seal is removed completely and the secondary seal on the lid used to secure the vial after use. For small volumes of isotope, manufacturers often recommend low-speed centrifugation to remove any solution from the top of the vial prior to opening.

Transport of radioactive materials between labs should be described in the local rules document. Alpha emitters should be transported in a secure plastic box, lined with absorbed paper. Beta emitters should be held in a Perspex box, while gamma emitters should be transported in a lead-lined container. Transport of materials by beyond the institution, for example to another place of business, is covered by specific legislation and professional carriers should be used. There are some exemptions, such as the transport of carbon-14-containing material with a total activity less than 370 MBq, which can be done in private vehicles.

Generally, those who handle radioactive materials and spend their time working in a supervised area are considered as *non-classified workers*, while those in controlled areas are *classified workers*. However, it is often the case that personnel move between labs; such movement must form part of an approved system of work designed to restrict a person's dose to less than three-tenths the dose equivalent. To ensure this is the case, both types of workers must record their exposure; in general:

(1) The activities of a non-classified worker involve regular monitoring of external dose through surface and air monitoring.

(2) With classified workers, personal monitoring is required to measure the individual dose. Film badges or *thermoluminescent dosimeters* (TLDs) are often used, but other techniques, such as thyroid scans may also be required. The personal monitoring records must be kept for 50 years after the last entry.

During work with radioisotopes, laboratory personnel should wear full PPE (Howie-style laboratory coat, safety glasses, and gloves). After completion of work and removal of gloves, a person should wash their hands and monitor them with a survey meter. If contamination persists, further washing is required. Similarly, a laboratory coat and safety glasses should be monitored, and washed accordingly. The decontamination of laboratory glassware and equipment depends on the nature of the radioactive material and its activity. Generally, glassware can be cleaned with water and an appropriate detergent (e.g., Decon-90). More rigorous cleaning agents, such as $K_2Cr_2O_7/H_2SO_4$, can be used provided there are no chemical incompatibilities. The use of plastic laboratory consumables (tubes, flasks, etc.) is recommended in some instances (such as in biomedical applications like tissue culture), although disposal of these increases the volume of radioactive waste.

For minor spills of radioactive material, any liquid should be absorbed by paper towels and the waste placed in plastic bags marked for disposal with an estimate of the total activity. The wash water should be directed down the laboratory sinks, taking care not to exceed the maximum permissible concentration. After the area has dried, wipe tests, or sweeps with a survey meter should be conducted. The cleaning process should be continued until the radiation has reached an appropriate level. For major incidents, the situation should be handled by the Radiation Safety Officer or other experience professional. If injuries occur, first aid should take priority — someone could potentially bleed to death long before the radiation levels exceed the allowed dose.

7.8 Detection of Radiation

Key Point: Radiation can be detected through its interaction with photographic film or by sensitive electronic detectors. These techniques are often qualitative.

One of the most basic means of detecting radiation is to use photographic film. At their simplest, photographic films, consist of small silver halide crystals (usually AgBr) embedded in gelatin. When light or ionizing radiation strikes the crystals, the salt becomes activated and forms a black silver grain. Unactivated silver halides are washed away during the development process, leaving the silver grains to form the negative image. This process is used in radiological protection, where personnel often wear *film badges*, which may appear as a blue plastic badge-holder, usually pinned to the lab coat at chest height. The assembly can accommodate different shielding materials depending on the type of radiation, which the user may be exposed to, plus a small piece of photographic film. The exposure of the film to radiation (up to a dose of 1 Gy) can be qualitatively estimated by development of the film. An alternative to film badges are the TLDs, which contain a phosphor (e.g., LiF), which remains in an excited state once exposed to radiation. Heat treatment of the film results in light emission with an intensity roughly proportional to the radiation dose (up to a maximum of 20 Gy).

The interaction of radiation with photographic film is also used in *autoradiography*, in which photographic film is exposed to a radioactive specimen. This usually involves radiolabeled materials separated by chromatography or electrophoresis, but could also include plant or animal tissue sections stained with a radioactive dye. When the film has been incubated with the sample, it must be developed in a dark room in much the same way as an "old fashioned" photograph. This involves developing the image with a reducing agent, quenching the development process in a stop bath, and then fixing the image using sodium thiosulfate. Modern autoradiography techniques now involve digital image capture using charged coupled devices. This has the obvious advantage of not requiring lengthy development steps and is amenable to automation.

7.9 Statistical Nature of Measuring Radiation Counts

Key Point: Quantitative measurements of radiation employ statistical concepts to take account of the random nature of radioactive decay. The main measure is the counting rate and its confidence interval.

When undertaking quantitative measurements of radioactivity, various descriptors are employed to define the actual measurement — for example, count rate, counts per minute, etc. However, when we recall the nuclear processes, which bring about radioactive decay, we remember that there is always a degree of uncertainty associated with the process. For a large number of atoms, the time interval when radioactive decay occurs can be calculated with reasonable accuracy. If the activity of a sample is measured, a large number of times and the data plotted as a frequency histogram, it would become obvious that the data centers on a mean value, which corresponds to the most probable value. We regard this as a *sample mean*, $\bar{X}$, which is an approximation of the *true mean*, μ. The certainty associated with the sample mean is described by its *standard deviation*, s, which describes how close the value of the sample mean is to the true mean.

In measurements of radioactivity, the standard deviation can be estimated from a single measured value, X, provided that the total number of counts exceeds one hundred (virtually always the case).

$$s = \sqrt{X} \tag{7.10}$$

Practically, radioactivity is measured by *counting rates*, R, which are the number of counts in a period of time:

$$R = \frac{X}{t} \tag{7.11}$$

Taking into account the high accuracy associated with measurements of time in modern instruments, we can ignore any deviations due to time and define the standard deviation of the counting rate as

$$s(r) = \frac{\sqrt{X}}{t} \tag{7.12}$$

To illustrate the point, suppose we recorded 1500 counts in 10 min; by Eq. (7.11) the counting rate would be 150 cpm. Similarly, following Eq. (7.12) the standard deviation would be 3.87 cpm. It is common practice to express this result using *confidence intervals* obtained from Eq. (7.13), where the value n is the interval factor (Table 7.5):

$$\frac{X}{t} \pm \frac{n\sqrt{X}}{t} \tag{7.13}$$

Table 7.5 Confidence Interval Factors

Interval	Confidence Level (%)
$\pm$ 0.67 s	50.0
$\pm$ s	68.3
$\pm$ 1.96 s	95.0
$\pm$ 2 s	95.5
$\pm$ 2.58 s	99.0

In our example, the 68.3% confidence interval (one standard deviation) would be taken as

$$\frac{1500}{10} \pm \frac{1\sqrt{1500}}{10} \text{ i.e., } 150 \pm 3.87 \text{ cpm}$$

Another commonplace means of expressing precision in measurements of radioactivity is the **relative deviation**, RD, which is similar to the idea of a coefficient of variation:

$$\text{RD} = \frac{\text{deviation value}}{\text{measured value}} \qquad (7.14)$$

Again, this can be considered as the RD of the counting rate, in which case Eq. (7.14) becomes

$$\text{RD} = \frac{s}{\left(\frac{X}{t}\right)} = \frac{\left(\frac{\sqrt{X}}{t}\right)}{\left(\frac{X}{t}\right)} = \frac{1}{\sqrt{X}} \qquad (7.15)$$

Recalling our earlier example, the RD would be evaluated as

$$\text{RD} = \frac{1}{\sqrt{1500}} = 0.026 \text{ (or 2.58\%)}$$

The significant of Eq. (7.15) should not be overlooked — it demonstrates that the relative uncertainty of a measurement is only dependent on the number of counts; it is independent of time.

Often, it is more useful to work in reverse. Suppose we wanted to ensure that a sufficient number of counts were obtained so that RD = 2% at the 95.5% confidence interval. In this case, we would have

$$0.02 = \frac{2\left(\frac{X}{t}\right)}{\left(\frac{X}{t}\right)} = \frac{2}{\sqrt{X}}$$

$$X = \left(\frac{1}{0.01}\right)^2 = 10000 \text{ counts}$$

Note that because we are considering the 95.5% confidence interval, we must multiply the standard deviation by 2 (*c.f.* Eq. (7.13)). The result of this calculation is worth remembering — if you collect 10000 counts, you will accrue a relative counting error of 2% at the 95.5% confidence interval.

It is often the case in practical measurements of radioactivity that the same measurement is made several times on the same sample, that is, replicates are used. In this case, we compute the sample mean and take the standard deviation for N replicates:

$$s(\bar{X}) = \frac{s}{\sqrt{N}} \tag{7.16}$$

It follows that the 68.3% confidence interval for the sample mean is given by

$$\bar{X} \pm \frac{s}{\sqrt{N}} \tag{7.17}$$

So far, we have neglected an important fact about measurements of radioactivity — the counting rate is actually the difference of the background counting rate and the measured counting rate. We describe this as the **net counting rate**. All of the preceding statistical treatment applies equally to both the background-counting rate and the measured counting rate. To evaluate the final experimental result, we must make use of the rules of propagation of error, which provide mathematical formulae to show how errors are handled in mathematical operations. If we consider the background count and time, X_b and t_b, and the measured count and time, X_m and t_m, we see that the net counting rate is

$$\left(\frac{X_m}{t_m}\right) - \left(\frac{X_b}{t_b}\right) \tag{7.18}$$

Which implies that the standard deviation and confidence interval of the net counting rate is given by Eqs. (7.19) and (7.20), respectively.

$$s(n) = \sqrt{\frac{R_m}{t_m} + \frac{R_b}{t_b}} \tag{7.19}$$

$$\frac{X_m}{t_m} - \frac{X_b}{t_b} \pm n\sqrt{\frac{X_m}{t_m} + \frac{X_b}{t_b}} \tag{7.20}$$

It should be obvious that a rigorous experimental approach would also take into account sources of error other than counting errors, such as those incurred through pipetting or weighing of the radioactive material. Determining the overall error of an experiment is well-covered in many analytical chemistry texts or more general works on laboratory statistics.

7.10 Counting Efficiency

Key Point: An instrument cannot detect all the species produced by a radioactive disintegration; the proportion detected is quantified as the counting efficiency.

The measurement of radioactivity by an instrument is associated with certain technical limitations — for example, not all the beta particles released by a sample of phosphorous-32 will be captured by the detection instrument. To take this into account, all instruments have an associated counting efficiency, which can be thought of as a form of calibration. If the radioactivity of a substance, A, is expressed in terms of the number of nuclei which disintegrate per unit of time — for example, one disintegration per second (equivalent to 1 Bq) — the measured radioactivity, n, of that substance corresponds to the number of disintegrations counted, expressed in counts per unit time, for example, counts per second. The measured radioactivity must be proportional to the actual radioactivity, that is,

$$A = \frac{n}{E} \quad \text{where } E \leq 1 \tag{7.21}$$

The proportionality constant, E, is known as the ***counting efficiency***, which is determined by the type of counting device used and is defined by

$$E = \frac{n}{A} = \frac{\text{counts per second}}{\text{disintegrations per second}} \tag{7.22}$$

Determination of the counting efficiency for a particular instrument is an important factor in radiation detection. It can be affected by factors, such as distance from the source, absorption or scattering of the radiation by air or the surface of the detection, and also by the detector efficiency (the proportion of radiation reaching the detector that causes a response). Most instruments are checked annually to verify their counting efficiency, especially those instruments, which can deteriorate with use (mainly gaseous ion detectors like the Geiger–Müller detector).

7.11 Gaseous Ion Detectors

Key Point: Gaseous ion detectors, such as the GM detector employ a gas-filled detection tube, which is ionized by incident radiation, creating a small current, which is proportional to the intensity of the radiation.

When an ionizing particle comes into contact with a gas, its electrostatic field can dislodge outer orbital electrons from the gas atoms/molecules, forming an ion pair: a negatively charged electron and a larger positively charged fragment ion. The minimum energy required for ion pair formation is referred to as the *ionization potential* and its magnitude depends on the nature of the gas and the energy of the incident radiation. In gaseous ion detectors, a tube is filled with an inert gas and fitted with two electrodes. When the charged particles are formed, they are attracted to their respective electrodes, where their charges are neutralized. Current then flows from a DC source through a high resistance circuit to restore the charge on the electrodes. The voltage drop across a resistor is measured, which provides an indication of the amount of current carried by the ions, which can then be related to the number of ions formed.

There are three main types of gaseous ion detector: ionization chambers, proportional counters, and Geiger–Müller tubes. While the basic principles of operation remain the same, each device operates at different voltages, which confer different measurement features. In *ionization chambers*, a relatively low voltage is applied between the electrodes, and the current is generated solely by the formation of the ion pairs. This allows detection of relatively high levels of radiation without the fill gas becoming depleted. In *proportional counters*, a slightly higher voltage operates across the electrodes, which has the effect of accelerating the electrons produced in ion pair formation, giving them sufficient energy to bring about further ion pair formations in what is known as the *Townsend avalanche*. This has the effect of increasing the sensitivity of the detector, such that a signal due to alpha or beta particles can be discriminated. The *Geiger–Müller detector* operates at relatively high voltages and employs both the Townsend avalanche and a second photoelectric process. The Geiger–Müller detector is virtually synonymous with practical radioactive measurements, largely because of its versatility and reliability. The references at the end of the book provide more complete descriptions of gas ionization detectors and their principles of operation.

7.12 Semiconductor Detectors

Key Point: Semiconductor detectors employ germanium diodes, which interact with radiation to create a minute electrical current, which can be detected.

In much the same way as ionizing radiation interacts with gases, it also interacts with semiconductors to produce electrons. In solid-state physics, **band theory** describes the electronic structure of solids in terms of allowed energy bands and forbidden band gaps. Those electrons, which are held tightly to the atom within a crystal lattice, constitute the valance band, which is contrasted by the mobile electrons found in the conduction band. These two regions are separated by the band gap, the size of which determines whether the solid is a conductor (small band gap), semiconductor (intermediate band gap), or insulator (large band gap). When ionizing radiation is incident upon a semiconductor, it excites electrons in the valance band, ejecting them leaving an electron hole. Under the influence of an electric field, the electron, and the hole move in opposite directions, creating a momentary electric current. Although the signal due to this current is incredibly weak, it can be amplified by an external circuit, leading to quantifiable detection of radiation.

Silicon is a classic semiconductor, which for the purposes of radiation detection is doped with traces of impurities, creating a **diode**. The diode has two regions: the n-type region (negative electrons migrate here) and the p-type region (which positive holes migrate toward). As the charged particles pass through banks of these diodes, the currents are detected and amplified. Germanium detectors are more often used in gamma ray spectroscopy rather than in general radiation detectors. In this technique, the energy spectrum of gamma rays can be recorded, which is a useful measurement in geology and astrophysics. To ensure an adequate signal-to-noise ratio, germanium detectors need to be cryogenically cooled in liquid nitrogen. This is because thermal energy is sufficient to promote formation of electron–hole pairs, which could be mistakenly interpreted as gamma emissions.

7.13 Scintillation Counters

Key Point: Scintillation counters employ chemical agents, which interact with radiation to produce fluorescence, which can be detected by photodetection systems.

When radiation interacts with certain substances, a scintillation process occurs, in which pulses of visible light are produced. Solid substances employed in scintillation counters include sodium iodide and anthracene, which can be used to measure emission of gamma rays and α-/β-particles, respectively. This is known as **solid scintillation counting** and for the most part this technique is employed in the detection of gamma rays.

The sodium iodide crystals used in gamma ray counters contain a small amount of thallium, which acts as an ***activation center*** from which the photons of light are released. As sodium iodide is particularly hygroscopic, it must be held in a sealed container, with the crystal wrapped in a magnesium oxide-coated aluminum film to protect it from extraneous light. The mechanism of light emission (***fluorescence***) in NaI(Tl) crystals is based on band theory, but differs from that used in semiconductor detection as there is no electric field present in detector. Consequently, when the electron–hole pairs are produced, they randomly migrate through the lattice until they recombine. As thallium has a smaller band gap than sodium iodide, the recombination is more likely to occur at this site, producing an excited thallium ion, which returns to its ground state by releasing a photon of visible light.

The scintillator is cylindrical in shape and has a cylindrical well bored through its center into which the sample is placed. This arrangement, plus the photomultiplier tube, is contained within lead shielding to protect it from external radiation. When the scintillation event occurs, the fluorescence is detected by a photocathode, which converts the light energy of the photons into electrons. The signal from the photocathode is amplified by the photomultiplier, which emits a series of electrical pulses. Each pulse represents a gamma ray photon, with the intensity of the fluorescence (and therefore electrical pulse) being proportional to the energy of the gamma ray. Most modern devices are referred to as ***multichannel counters*** in which a high number of samples can be processed. These have found particular application in the biological setting where radioimmunoassay and immunoradiometric assays are used. These are often high throughput processes where automation and accuracy are essential (e.g., in the measurement of hormones in human serum).

Detection of radiation from pure beta emitters (also common in biological settings) generally involves ***liquid scintillation counting*** in which the beta particles interact with a ***scintillation cocktail***. A scintillation cocktail is a solution of a fluorescent compound (the ***fluor***) dissolved to a very low concentration in an organic solvent (usually methylbenzene). When the beta particles strike the scintillation cocktail, their energy is initially transferred to the solvent, which after a number of transfers, eventually arrives at a fluor molecule. The excitation and relaxation of the fluor produces the fluorescence, which is detected by a photomultiplier tube. Some common fluors are shown in Table 7.6.

Table 7.6 Common Fluors in Liquid Scintillation Counting

Fluor	Structure
2,5-Diphenyloxazole (PPO)	
2-(4-*tert*-Butylphenyl)-5-(4-biphenyl)-1,3,4-oxadizole (Butyl-PBD)	
2,5-Bis-2-(5-tertbutylbenzoxazolyl)-thiophene (BBOT)	
p-Bis-(O-methylstyryl)-benzene (Bis-MSB)	
1,4-Bis-2-(5-phenyloxazoyl)-benzene (POPOP)	
1,4-Bis-2-(4-methyl-5-phenyloxazoyl)-benzene (Dimethyl-POPOP)	

In practice, scintillation cocktails contain two different fluorescent compounds, with the one present at higher concentrations known as the **primary fluor**. The second compound, the **secondary fluor**, acts as a type of filter, changing the wavelength of light emitted by the scintillation process. The inclusion of a secondary fluor is usually because the wavelengths of light emitted by the primary fluor are too short to be detected by the photomultiplier tube. For example, 2,5-diphenyloxazole has a maximum fluorescence at 365 nm, but the inclusion of 1,4-*bis*(5-phenyl-2-oxazolyl)benzene extends the wavelength to 429 nm. This greatly improves the quantum efficiency of the measurement — that is, it produces more detectable photons and a signal with a higher signal-to-noise ratio.

Chapter Summary

- Radioisotopes used in industrial or other settings are classified as sealed or unsealed sources. The unsealed sources are more commonly used in research or analytical laboratories and these carry higher levels of risk to the user.

- Exposure of an individual to radiation is ultimately measured by the effective dose, which takes into account the specific properties of the radiation and the tissue, which are exposed. The SI unit of effective dose is the sievert, Sv.
- External exposure to radiation can be reduced by appropriate shielding, which can be determined by using the HVL for different materials. Internal exposure to radiation is much harder to reduce once it has occurred, particularly if the effective half-life of the radionuclide is long.
- Levels of exposure should be minimized if legislative requirements are followed. Key legislation in the UK, IRR99, and RSA93, sets limits on exposure for various individuals, describes how radioactive substances should be stored and handled and identifies the key features of local rules.
- Radiation can be qualitatively detected by photographic film and this forms the basis of personal monitoring devices, or laboratory techniques such as autoradiography. Quantitative detection of radiation obeys statistical laws and uses three main techniques: gaseous ion detectors, semiconductor detectors or scintillation detectors.

Review Questions

(1) Suggest why many laboratories prefer to label compounds with deuterium rather than tritium.

(2) Define "stochastic" in the context of radiation protection and highlight any problems associated with mitigating so-called stochastic effects.

(3) Suggest the meaning of the term "critical organ."

(4) Calculate the exposure rate from an unshielded iodine-131 source at a distance of 3 m if its activity is 5×10^{10} dpm.

(5) Determine the minimum number of lead bricks (2 cm thick; HVL of 1.2 cm) needed to reduce the exposure at a distance of 45 cm from a 0.5 Ci cobalt-60 course to 10 mR/h.

(6) If a designated worker is handling a 370 MBq caesium-137 source at a distance of 50 cm from their head, estimate the total dose received by their eyes over a 30 min period.

(7) An average 70 kg man received an internal dose of 1.0 mCi of phosphorous-32. Calculate the initial dose rate assuming an average energy per disintegration of 700 keV.

(8) The following data were collected in an experiment utilizing liquid scintillation counting: 3775, 4194, 3697, 4328, and 3986. Evaluate the sample mean, standard deviation and 95% confidence interval for this data.

(9) Demonstrate by a suitable means that a measured count of 67000 is needed to achieve a 1% deviation at the 99% confidence interval.

(10) Suggest why the common fluors shown in Table 7.6 are based on a conjugated aromatic skeleton.

Chapter 8

The Nucleus, Spectroscopy, and Spectrometry

"A wire telegraph is a kind of a very, very long cat. You pull his tail in New York and his head is meowing in Los Angeles. Radio operates exactly the same way: you send signals here, they receive them there."

A. Einstein

Spectroscopy and spectrometry are the mainstays of modern chemical analysis, with nuclear properties having considerable impact on the data obtained. On completion of this chapter and the associated questions you should:

- Have an overview of divisions of spectroscopy and their foundations in quantum mechanics.
- Understand the role of the nucleus in spectroscopy through reference to spectroscopic theory.
- Be able to identify the key features of mass spectrometry and its application in the determination of isotopic ratios.

8.1 Principles of Spectroscopy

Key Point: Spectroscopy is the study of the interaction of electromagnetic radiation with matter in accordance to the Planck–Einstein relation, $\Delta E = h\nu$.

Spectroscopy studies the interaction of electromagnetic radiation with matter; spectrometry is based on different principles (Section 8.5). When we

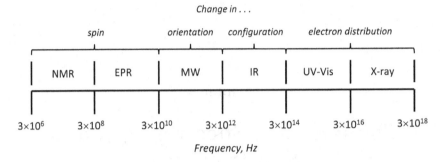

Figure 8.1 Regions of the electromagnetic spectrum used in spectroscopy.

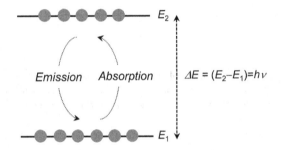

Figure 8.2 Energy level transitions. Both energy levels are populated at $T > 0$ K according to the Boltzmann distribution, with the lower energy level, E_1, maintaining a slight excess. When EmR of the correct frequency is absorbed, a transition from $E_1 \rightarrow E_2$ occurs. The return to ground state ($E_2 \rightarrow E_1$) is associated with the emission of a photon.

consider the breadth of the electromagnetic spectrum, it's hardly surprising that a large number of spectroscopic techniques have been developed to investigate the structure of atoms and molecules. A summary of the main forms of spectroscopy and how they relate to the electromagnetic spectrum is given in Figure 8.1. While it is immediately obvious that some forms of spectroscopy, such as nuclear magnetic resonance (NMR), will have a strong dependence on nuclear properties, for other techniques, particularly those examining electron behavior, this dependence may not be so clear.

Any atom or molecule possesses different forms of energy which are constrained to exist in discrete amounts (quanta). Consider the simple scenario represented by Figure 8.2 in which two energy states, E_1 and E_2, are arbitrary energy levels in an atom. It is possible for an electron to move between the energy levels provided that a suitable amount of energy, ΔE, can be absorbed or emitted. It was Planck who suggested that this energy could be in the form of electromagnetic radiation, EmR, and would have a magnitude given by the familiar relationship:

$$\Delta E = E_2 - E_1 = h\nu \qquad (8.1)$$

If we were to direct EmR at the atom, a portion of its energy would be absorbed to mediate the transition from E_1 to E_2 and the overall intensity of the EmR would decrease. If a suitable detector were positioned to monitor the transmitted EmR, its output would show that only energy of specific frequencies was absorbed, producing an **absorption spectrum**. It is also possible that a transition from E_2 to E_1 may occur, releasing a photon of EmR and producing an **emission spectrum** which is complimentary to the absorption spectrum.

The signal recorded in spectroscopic techniques is a consequence of **resonance** — when the energy supplied or emitted matches the energy of the system under study. The spectra produced from this signal consist of a series of peaks which represent the range of frequencies involved in a transition. The smallest possible range of frequencies for a transition produces the **natural linewidth**, Γ ("gamma"), which arises from the uncertainty associated with a specific energy level:

$$\delta E \delta t \approx \frac{\hbar}{2} \qquad (8.2)$$

It follows that if the energy level has uncertainty, then the transition between energy levels must also have uncertainty and the precise frequency of a transition between energy levels can never be known. We use the **width at half height** to measure natural linewidth (Figure 8.3), or uncertainty in frequency, which can be reasonably expressed as

$$\Gamma = \delta \nu = \frac{1}{2\pi \delta t} \qquad (8.3)$$

In addition to this, the instruments used to measure absorption/emission spectra have limited resolution which means that the linewidth of the observed peaks spans a frequency range.

The linewidth of a peak is influenced primarily by two phenomena: the Doppler effect and lifetime broadening. The **Doppler effect** was first recognized with sound by Johann Doppler (1803–1853) who reported that a system under study will be observed to undergo a change in frequency relative to a static observer. From a spectroscopy perspective, since atoms and molecules are in constant random motion, the shift in frequency will be in both directions, leading to a broadening of the peak.

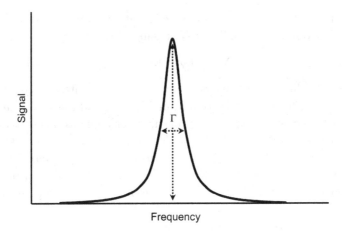

Figure 8.3 Linewidth. The width at half height provides a measure of the spectral linewidth, which in turn is an indication of the uncertainty associated with a transition between energy levels.

The second mechanism, **lifetime broadening**, is significant even at very low temperatures as its origins are quantum mechanical. When a system is in its excited state, the energy dissipates in an exponential fashion according to its **lifetime**, τ. If the system had an infinite lifetime, its energy could be precisely determined; however, because the uncertainty principle precludes this, the excited state cannot have a precisely defined energy and will follow Eq. (8.3) (where $\delta t = \tau$). It follows that as the spread in energy is inversely related to the lifetime, then the shorter the lifetime, the broader the spectral line. For an electronic transition with a lifetime of 10^{-8} s, the uncertainty (broadening) of the peak would be around 10^8 Hz. Given that the frequencies associated with this type of transition are 10^{14}–10^{16} Hz, the overall effect of lifetime broadening would be relatively small. However, in other types of transition, the impact of lifetime broadening can be more significant.

The final feature of a spectroscopic peak is the **intensity** which is an indication of the amount of energy absorbed/emitted by the system. This has three main contributing factors. The first is the **transition probability**, which is governed by selection rules. These describe allowed or forbidden transitions and can be used to predict the appearance of peaks on spectra. The second factor is the population of energy states which is governed by the **Boltzmann distribution**:

$$\frac{N_{\text{upper}}}{N_{\text{lower}}} = \mathrm{e}^{-\Delta E/kT} \qquad (8.4)$$

At equilibrium when $T > 0\,K$, $N_{lower} > N_{upper}$ with thermal collisions being responsible for population of the upper energy level. The net rate of absorption of EmR is proportional to the difference in energy states and the intensity of absorption is the product of this and the energy per photon. Finally, the **pathlength** of the sample (essentially the "thickness" of the sample) is directly proportional to the amount of energy absorbed. This latter factor is particularly important for transitions in the UV–Vis range.

8.2 Magnetic Resonance Spectroscopy

Key Point: Moving charged particles produce a magnetic moment which interacts with an external magnetic field, splitting the energy levels. This is known as the Zeeman effect.

In the late 1930s, it was observed that certain nuclei absorb radiofrequency radiation when placed in a magnetic field. When the magnetic field strength was held constant and the frequencies of radiation varied, nuclei with odd numbered spin resonated. A few years later, a similar effect was observed for species with an unpaired electron. This phenomenon became known as magnetic resonance from which NMR spectroscopy and electron paramagnetic resonance (EPR) spectroscopy were developed.

A key feature of magnetic resonance techniques is the effect of an applied magnetic field on transitions between energy levels. Earlier, in Chapter 3, we introduced the idea of angular momentum and that a spinning, charged particle would produce a magnetic moment, μ. When a static magnetic field, B_0, is applied the energy of the particle is shifted (Figure 8.4).

$$\Delta E = \mu B_0 \tag{8.5}$$

This splitting of the energy level is known as the **Zeeman effect**, after Dutch physicist Pieter Zeeman (1865–1943). In practice, the Zeeman splitting shown in Figure 8.4 is rarely observed except at very low resolution. Instead several lines are often observed, which are due to the coupling of the magnetic moments of orbital spin (above) with the magnetic moment associated with the intrinsic spin of the particle. This is sometimes referred to as the **anomalous Zeeman effect**.

8.2.1 Nuclear magnetic resonance spectroscopy

Key Point: Nuclei in a magnetic field are split into two populations; transitions between the two populations involve absorption/emission of radiofrequency radiation, producing a signal.

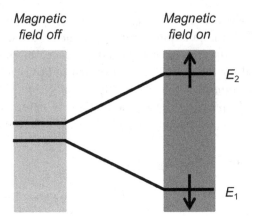

Figure 8.4 The Zeeman effect. Energy levels in their ground state are split by an external magnetic field. The magnitude of the energy difference is given by the Planck–Einstein equation, $\Delta E = (E_2 - E_1) = h\nu$.

Table 8.1 Properties of Common Nuclides in NMR

Nuclide	Abundance, %	Spin	$\gamma(\times 10^5)\mathrm{T}^{-1}\mathrm{s}^{-1}$
^{1}H	99.98	1/2	26.75
^{2}H	0.02	1	4.11
^{13}C	1.11	1/2	6.73
^{14}N	99.64	1	1.93
^{19}F	100	1/2	25.18
^{31}P	100	1/2	10.84

Nuclei with odd-numbered spin will exert a magnetic moment given by

$$\mu = \gamma I \qquad (8.6)$$

where γ is the ***magnetogyric ratio*** (unique to a particular nucleus; Table 8.1) and I is the nuclear spin quantum number. Degenerate nuclei possessing this magnetic moment will interact with an external magnetic field, lifting their degeneracy and creating an energy difference. The magnitude of the energy difference is given by

$$\Delta E = h\nu = \gamma \hbar B_0 \qquad (8.7)$$

The population of the two energy levels is heavily dependent on temperature (according to the Boltzmann distribution); at 298 K, thermal excitation has raised many nuclei to the higher energy state, although a slight excess still remain in the lower energy level. If nuclei are supplied with energy in the form of radiofrequency radiation, they flip between the high and low energy states. The frequency at which this occurs is known as the **Larmour frequency**, which will be given by rearrangement of Eq. (8.7):

$$\nu = \frac{\gamma \hbar B_0}{h} = \frac{\gamma B_0}{2\pi} \tag{8.8}$$

The Larmour frequency is directly proportional to the strength of the magnetic field.

When the radiofrequency radiation is applied, the distribution of nuclei between the states equalizes, implying that no further absorption of energy should occur. This contradicts the observed magnetic resonance effect, which means that a mechanism exists to restore the unequal distribution of nuclei. This mechanism is referred to as **relaxation** of which there are two kinds:

(1) Spin-lattice relaxation: Excited nuclei interact with the magnetic field of adjacent atoms, bringing the excited nuclei back to their ground state and releasing the energy as heat (enthalpic relaxation). This process maintains the excess of nuclei in the lower energy state and is measured by the time T_1.

(2) Spin–spin relaxation: Dipole–dipole interactions between excited nuclei and ground state nuclei lead to an exchange of energy and an increase in randomness (entropic relaxation). The relaxation time for this process is designated T_2 and gives the half-height of the spectral peak. Rapid motion increases T_2 which produces a narrow spectral peak.

Relaxation times are particularly important in resolving healthy and diseased tissue by MRI. Contrast agents which enhance the contrast of the image are paramagnetic compounds which typically shorten the relaxation time of protons with long T_2.

Even for the simplest nuclei, the surrounding electron density reduces the effect of the applied magnetic field through **shielding**. Nuclei of the same isotope present in different molecules will have different shielding constants and therefore have different resonant frequencies. This produces the **chemical shift**, δ, which is used to identify different functional groups. The chemical shift is an empirical quantity which relates the resonant frequency of the nucleus under study to that of a reference standard, ν_o:

$$\delta = \frac{\nu - \nu_o}{\nu_o} \times 10^6 \tag{8.9}$$

The most common reference standard is tetramethylsilane, $Si(CH_3)_4$, usually referred to as TMS. Nuclei with a small shielding constant and large chemical shift are said to be **deshielded** and the peaks for these nuclei are described as being downfield.

The chemical shift of a nucleus can be influenced by a variety of factors. A commonly encountered example in ^{1}H-NMR is the effect of electronegative atoms on protons which moves the chemical shift downfield for example, the methyl groups in CH_3CH_3 have $\delta = 0.96$ while in CH_3Cl $\delta = 3.05$. A nucleus can also be shielded or deshielded by the magnetic properties of adjacent groups (known as **magnetic anisotropy**). The most common example of this is the effect of the delocalized electrons in benzene which can produce very large shielding effects for protons above the plane of the ring, and smaller deshielding of protons to the side of the ring.

The magnetic moments of adjacent non-equivalent nuclei interact to produce the fine structure of the NMR spectrum. This is termed **spin–spin coupling** and leads to the splitting of a peak according to the multiplicity rule:

$$M = 2nI + 1 \qquad (8.10)$$

where n is the number of neighboring nuclei. For ^{1}H-NMR, this is reduced to the $n + 1$ rule: if a set of hydrogen atoms is adjacent to n non-equivalent hydrogens, its peak will be split into $n + 1$ ways. The intensities of the peaks are given by the binomial distribution (Pascal's triangle). The **coupling constant** J measures the splitting between two different nuclei in a given molecule. In ^{1}H-NMR, protons which are two bonds apart (germinal coupling) have $J = 2{-}15\,\mathrm{Hz}$, while those three bonds apart (vicinal coupling) have $J = 2{-}20\,\mathrm{Hz}$. Coupling with more remote protons is possible if the geometry of the molecule right, or pi-bonding systems are involved.

In order to reduce artifactual signals in ^{1}H-NMR, deuterated solvents such as deuterium oxide, deuterochloroform or dimethylsulfoxide-d^6 are used to dissolve/dilute samples (^{2}H is NMR silent). Deuterium oxide is also useful in the detection of acidic hydrogen atoms by ^{1}H-NMR as it undergoes a rapid exchange with these protons in a process known as **deuterium exchange** (the "D_2O shake"):

$$ROH + D_2O \rightleftharpoons ROD + DHO$$

The signal originally due to the hydroxyl group will disappear, as will any spin–spin splitting caused by the protons. Other groups, such as amide protons and protons adjacent to carbonyl groups (α-hydrogens) also undergo deuterium exchange (Figure 8.5). Deuterium and other nuclei with

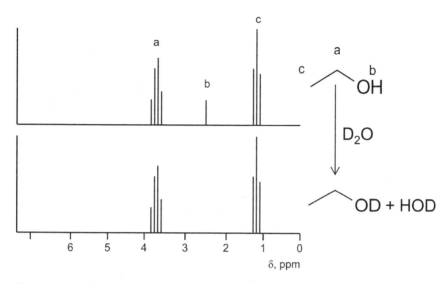

Figure 8.5 Deuterium exchange in ethanol. The top ^{1}H-NMR for ethanol shows the peak assignment for each group. After shaking with D_2O, the hydroxyl peak (b) disappears.

$I > 1/2$ possess a **quadrupole moment** due to an uneven distribution of charge. This causes broadening of the peak due to a longer T_1 and can significantly affect resolution of splitting patterns.

Aside from ^{1}H-NMR, ^{13}C-NMR is the most commonly employed NMR technique. Carbon-13 has a smaller nuclear magnetic moment than of a proton, meaning that the NMR signal is weaker and more difficult to observe. To compensate for this, larger quantities of sample are often used and a larger number of scans are taken to compile the spectrum. Using the data in Table 8.1 and Eq. (8.8) we can show that the resonant frequency of a carbon-13 nucleus in a 10 T magnetic field is much lower than that of a proton (*ca.* 425 MHz):

$$\nu = \frac{(6.7 \times 10^7) \times 10}{2\pi} = 10.7 \times 10^7 \, \text{Hz} = 107 \, \text{MHz}$$

The range of chemical shifts of nuclei in ^{13}C-NMR is much broader than in ^{1}H-NMR which virtually eliminates overlapping.

Carbon-13 is regarded as a dilute-spin species — it is unlikely that more than one carbon-13 nucleus will be found in any given sample. This means that splitting due to **homonuclear coupling** is very unlikely to occur. Conversely, protons are an abundant-spin species and they can result in a very complex spectrum due to **heteronuclear coupling** of one carbon-13 nucleus with many protons. To avoid this difficulty, a technique known as **proton decoupling** is used. In this approach, the associated protons are exposed to a high-intensity radiofrequency pulse set to their resonant frequency. As the attached protons flip between the two energy states, the carbon-13 nucleus experiences an average magnetic field from the protons and no coupling takes place. However, the decoupled proton can still have an effect on the carbon-13 nucleus through the **nuclear Overhauser effect**. This arises from direct magnetic coupling of a decoupled proton and the carbon-13 nucleus and increases the number of nuclei in the lower energy level of the carbon-13. This has the effect of increasing the signal (and therefore sensitivity) of ^{13}C-NMR, but removes the proportionality between peak areas and number of nuclei.

Other than ^{1}H- and ^{13}C-NMR, most work is done with $I = 1/2$ nuclei, such as ^{19}F, ^{31}P and increasingly ^{77}Se, ^{119}Sn, and ^{199}Hg. The study of nuclei other than ^{1}H and ^{13}C is referred to as **multinuclear NMR** which is an increasingly utilized technique, especially in bioinorganic and organometallic chemistry. Even though a nucleus may have non-integer spin, it must also be present at a reasonable abundance in order for NMR to be a practical technique. The fluorine-19 isotope has 100% abundance and is comparable to the proton in terms of sensitivity. As it is highly electronegative, the chemical shifts in ^{19}F-NMR are distributed over a much greater range. This has its advantages: compounds which are structural homologues can often be distinguished by ^{19}F-NMR, especially since fluorine has only a small effect on stereochemistry (due to its smaller van der Waals radius). Similarly, ^{31}P-NMR exhibits sharp peaks and has been used extensively in the study of biological phosphorous chemistry. For example, the lanthanides are known the form paramagnetic complexes with nucleic acids and have been used to provide contrast in magnetic resonance imaging. These elements also show some promise at chemotherapeutic agents, as their high Lewis acidity can bring about the hydrolysis of phosphodiester bonds in DNA.

8.2.2 Electron paramagnetic resonance spectroscopy

Key Point: A single unpaired magnetic field occupies one of two energy states. Transitions between the two states are mediated by absorption/emission of microwave radiation.

In EPR spectroscopy, we consider a single unpaired electron as having degenerate random spin. When it is exposed to an external magnetic field, the degeneracy of the electron spin is lifted, forcing it to adopt either a higher or lower energy state. The difference in energy is related to the properties of the system by

$$\Delta E = h\nu = g\mu_{B}B_{0} \tag{8.11}$$

The coefficient g is known as the g-factor which for a free electron has the value 2.0023 and the constant μ_{B} is the Bohr magneton ($9.274 \times 10^{-24}\,\mathrm{J\cdot T^{-1}}$). It is a relatively simple matter to show that the frequency of EmR corresponding to the energy difference is in the microwave region, specifically the X-band (8–10.9 GHz).

In a typical EPR experiment, the frequency of microwave radiation is held constant and the magnetic field is varied until resonance is achieved. The intensity of the signal is proportional to the change in sample microwave absorption. The intensity of the curve is proportional to the amount of material present (sensitive to about $10^{-12}\,\mathrm{mol}$) and the position of the EPR corresponds to the resonance condition, which is reported as an observed g-value (Figure 8.6). The characteristic feature of an EPR spectrum is the hyperfine structure, which arises from electron–nucleus interactions and acts as a form of fingerprint for a particular species.

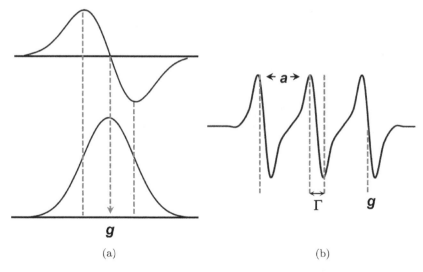

(a)   (b)

Figure 8.6 EPR signals. (a) Generalized EPR signal (lower) with first derivative curve (upper). (b) First derivative curve showing measurement of g-factor, g, and the hyperfine coupling constant, a.

Isotopes with even–even nuclei (^{12}C, ^{28}Si, ^{56}Fe etc.) are EPR silent; those with odd–even nuclei (^{2}H, ^{10}B, ^{14}N etc.) or even–odd nuclei (^{1}H, ^{13}C, ^{19}F, ^{55}Mn etc.) produce EPR signals. As in NMR, a single nucleus with $I = 1/2$ will split each energy level into two, giving two energy transitions (absorptions) whose difference is equal to the **hyperfine coupling constant**, a, which is characteristic for a particular species. If only one nucleus is interacting with the unpaired electron, all lines will have equal intensity; if multiple nuclei are interacting, the intensities follow a binomial distribution.

We can see these effects by comparing two distinct species, vanadium acetylacetonate and the pyrazine anion. The vanadium nucleus in VO(acac)$_2$ has $I = 7/2$. We can predict the number of lines using the multiplicity rule which gives $2(1)(7/2) + 1 = 8$ lines of equal intensity (only one equivalent nuclei) as shown in Figure 8.7. With the pyrazine anion, the unpaired electron is delocalized over the ring and couples initially with the two nitrogen atoms to give $2(2)(1) + 1 = 5$ lines with intensities of 1:2:3:2:1 (a quintet). However, it then couples with the hydrogen atoms to split the original quintet into $2(4)(1/2) + 1 = 5$ lines.

The investigation of direct hydrogen/deuterium substitutions by EPR spectroscopy is fairly limited as nuclear mass has little direct impact on the EPR spectrum. EPR spectroscopy can be used to determine isotopic abundance in paramagnetic compounds. For example, potassium nitrosodisulfonate, also known as Fremy's salt, K$_4$[(SO$_3$)$_2$(NO)]$_2$, is a

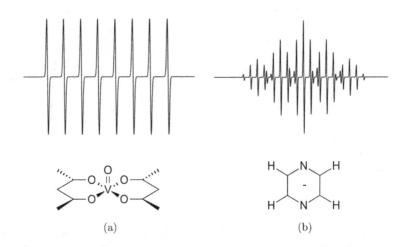

(a) (b)

Figure 8.7 EPR signals for vanadium acetylacetonate (a) and pyrazine anion (b).

well-known paramagnetic compound used as a source of nitrosyl radicals. The dianion has a distinctive triplet EPR signal which can be used for calibration of instruments. It has been shown that it is possible to measure hyperfine splitting due to ^{17}O, ^{33}S, and ^{15}N isotopes present in the salt. Taking into account the low natural abundance of these isotopes (0.037, 0.74 and 0.365%, respectively), EPR spectroscopy would seem to have the potential to be a sensitive means of determining isotopic abundance in samples.

8.3 Mössbauer Spectroscopy

Key Point: Gamma rays emitted by specific source nuclei are absorbed by their corresponding sample nuclei. Subtle shifts in resonant frequency provide information about oxidation state.

Mössbauer spectroscopy, named after Rudolf Mössbauer (1929–2011), is a form of gamma ray spectroscopy which uses the resonant absorption of gamma rays by atomic nuclei to identify chemical and physical properties of the substance. As with other spectroscopic techniques, the resonance condition requires that the energy of the incident radiation must match the difference in nuclear energy levels. Although gamma ray spectroscopy has been known for some years, it was Mössbauer in 1957 who realized that for the technique to work, the source of the gamma rays had to be fixed in a lattice to minimize recoil effects.

An example of the Mössbauer effect can be seen with iron-57 nuclei which interact with gamma rays released from the decay of cobalt-57. The unstable cobalt-57 nucleus ($t_{1/2} = 270\,\text{days}$) first absorbs an inner shell electron to produce an excited iron-57 nucleus, an electron neutrino and a gamma ray photon:

$$^{57}_{27}\text{Co} + e^- \rightarrow {}^{57}_{26}\text{Fe}^* + \nu_e + \gamma\,(122\,\text{keV})$$

The excited iron-57 nucleus then rapidly decays ($t_{1/2} = 2 \times 10^{-7}\,\text{s}$) to its ground state releasing a 14.5 keV gamma ray photon, which is the resonant photon for the iron-57 sample:

$$^{57}_{26}\text{Fe}^* \rightarrow {}^{57}_{26}\text{Fe} + \gamma\,(14.5\,\text{keV})$$

Due to the uncertainty associated with all quantum mechanical systems, the frequency of the gamma ray photon released will also carry some uncertainty. We can show this by first calculating the lifetime of the excited iron-57:

$$\tau = 1.443 \times 2 \times 10^{-7}\,\text{s} = 2.88 \times 10^{-7}\,\text{s}$$

This is relatively long and would be expected to be associated with a small uncertainty in energy:

$$\delta E = \frac{\hbar}{\tau} = \frac{1.054 \times 10^{-34}\,\text{J} \cdot \text{s}}{2.88 \times 10^{-7}\,\text{s}} = 3.66 \times 10^{-28}\,\text{J}$$

We can easily convert this result into an uncertainty in frequency using the Planck equation:

$$\delta\nu = \frac{h}{\delta E} = \frac{3.66 \times 10^{-28}\,\text{J}}{6.626 \times 10^{-34}\,\text{J} \cdot \text{s}} \approx 5 \times 10^{5}\,\text{s}^{-1}$$

which is small in comparison to the frequency of a 14.5 keV gamma ray photon (3.50×10^{18} Hz). In fact, the relative line width, $\delta\nu/\nu \approx 10^{-13}$ is much smaller than that found in many other spectroscopic techniques.

Despite the small line width, early attempts at gamma ray spectroscopy were plagued with inaccuracies due to Doppler broadening. When the photon is ejected from the nucleus, the nucleus will experience recoil due to the momentum of the gamma ray photon. For an iron-56 nucleus, the recoil velocity would be

$$u = \frac{h\nu}{mc} = \frac{(6.626 \times 10^{-34}\,\text{J} \cdot \text{s}) \times (3.5 \times 10^{18}\,\text{s}^{-1})}{(9.47 \times 10^{-26}\,\text{kg}) \times (2.99 \times 10^{8}\,\text{m} \cdot \text{s}^{-1})} = 81.9\,\text{m} \cdot \text{s}^{-1}$$

The frequency shift is given by Eq. (8.12) which when applied to the case of the iron-57 nucleus gives an approximate value of 10^{13}

$$\Delta\nu = \frac{\nu u}{c} = \frac{(3.5 \times 10^{18}\,\text{s}^{-1})(87.1\,\text{m} \cdot \text{s}^{-1})}{2.99 \times 10^{8}\,\text{m} \cdot \text{s}^{-1}} \approx 1 \times 10^{12}\,\text{s}^{-1} \qquad (8.12)$$

This value is relatively small when compared to the emission frequency of the gamma ray (*ca.* 10^{18} Hz) but is much larger than the line width (*ca.* 10^{5} Hz). Mössbauer discovered that if the emitter was held tightly in a cooled crystalline lattice, the recoil energy could be dissipated, reducing the frequency shift.

A nucleus in different chemical surroundings from the source will not absorb gamma rays at exactly the same frequency. To compensate for this, the source in a Mössbauer spectrometer is mounted on an oscillating drive to vary the frequency (by way of the Doppler effect) and the resultant spectrum is usually a graph of gamma counts per second versus centimeters per second. The fact that the chemical environment affects the absorption

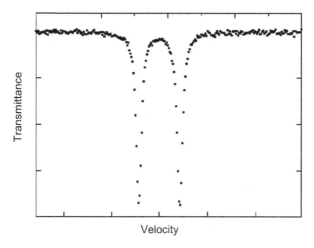

Figure 8.8 Mössbauer spectrum for nitroprusside ion. The two peaks are due to half integer spin of the iron nucleus.

of gamma rays produces a chemical shift (isomer shift). The main factor contributing to the chemical shift is the s-orbital electron density at the nucleus. The Mössbauer spectrum therefore gives an indication of the bonding about the nucleus. For example, when tin(II) loses its $5s^2$ electrons, the chemical shift changes by $3.7\,\text{mm}\cdot\text{s}^{-1}$.

Many Mössbauer isotopes have non-zero nuclear spin and therefore possess a quadrupole moment. For example, the excited iron-57 nucleus has $I = 3/2$ which produces a quadrupole moment and splits the nuclear energy levels into two substates. Such an effect can produce useful spectroscopic detail, such as in the Mössbauer spectrum of the nitroprusside ion, $[\text{Fe(CN)}_5\text{NO}]^{2-}$ (Figure 8.8). Hyperfine splitting also occurs in Mössbauer spectroscopy, where the magnetic field of surrounding nuclei split the energy levels into $2I + 1$ substates. For iron-57, the $I = 3/2$ state will be split into four substates and the $I = 1/2$ is split into two substates, giving a total of six potential transitions.

The prevalence of iron in biological compounds has made Mössbauer spectroscopy a popular technique with bioinorganic chemists. For example, the iron–sulfur protein ferrodoxin is found in the chloroplasts of green algae where it is believed to play a role in photosynthesis by acting as an electron acceptor/donor. The Mössbauer spectrum of ferrodoxin (Figure 8.9) shows that in the oxidized state, both the iron atoms are present as iron(III), while in the reduced state, only one of the iron atoms is present as iron(II).

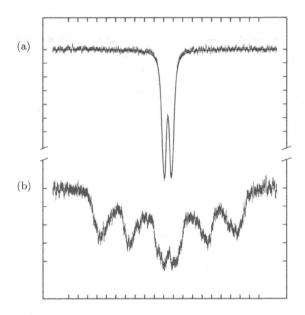

Figure 8.9 Mössbauer spectrum of ferrodoxin. (a) oxidized; (b) reduced.

Minaturized Mössbauer spectrometers (MIMOS II) were selected by NASA for inclusion on the two Mars rovers, Spirit and Opportunity, deployed in January 2004. The MIMOS II instruments, together with a X-ray fluorescence spectrometer and a microscopic imager, have been gathering geological data on the Martian surface. Some of the most interesting data has been gathered by Opportunity, which was released on the Meridiani Planum region, where geologists suspected there may be evidence of water in the geological history of Mars. Data gathered by MIMOS II indicated the presence of the mineral jarosite, $KFe_3(OH)_6(SO_4)_2$, which can only form in the presence of water.

8.4 Infrared Spectroscopy

Key Point: Different covalent bonds absorb different frequencies of infrared radiation according to the harmonic oscillator model at low energies or the anharmonic oscillator at higher frequencies.

Infrared spectroscopy is a versatile method commonly used for the identification of compounds. It is based on the observation that infrared radiation causes covalent bonds to vibrate, with the amount of energy absorbed related to the type of covalent bond (C=C, C−N, O−H, etc.). Much of the fundamental theory of infrared spectroscopy considers only diatomic molecules which are studied in the gas phase to allow relatively free rotation of the molecule. In fact, infrared spectra are a result of transitions between vibrational and rotational energy levels, although we will focus mainly on the contribution of the former.

Consider the formation of a diatomic molecule such as hydrogen chloride. Setting aside the creation of molecular orbitals, the formation of a covalent bond requires a balancing of two opposing forces: repulsion between the two positive nuclei and two negative electron clouds, and attraction between the nucleus of one atom and the electrons of the other. At some point, a balance is achieved and the two atoms settle at a distance equal to the bond length where the total energy is at a minimum. We can compress or extend this bond by changing the energy of the system and model the behavior of the bond using the *harmonic oscillator* (*c.f.* Chapter 1).

The force required to extend the bond from its equilibrium position to some other position ($r - r_{eq} = x$) is related to a force constant, k, which measures the stiffness of the bond:

$$F = -kx \tag{8.13}$$

As we are dealing with a conservative system, force can be related to the potential energy of the system:

$$F = -\frac{dV}{dx} \tag{8.14}$$

which means that the potential energy of the covalent bond is given by the integral:

$$V(x) = \int \frac{dV}{dx} dx = -\int F \cdot dx = \frac{kx^2}{2}$$

$$\therefore V(x) = \frac{1}{2}kx^2 \tag{8.15}$$

Equation (8.15) is that of a parabola, which is shown with the actual potential energy curve for a covalent bond in Figure 8.10. From this, we see that the harmonic oscillator is a good approximation for the behavior of a covalent bond at low energies. At higher energies, an *anharmonic oscillator* model is used which is based on the empirically derived Morse potential.

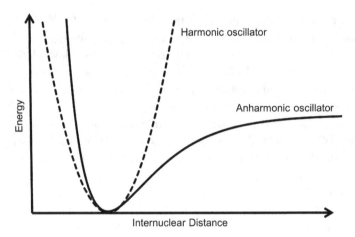

Figure 8.10 The harmonic oscillator versus the anharmonic oscillator models. At lower energy levels, both models follow a similar form, but as energy increases, the models diverge with the anharmonic oscillator model providing a more realistic interpretation.

The vibrational energy of the covalent bond is quantized using the vibrational quantum number v and the allowed energies are calculated from the Schrödinger equation. For the simple harmonic oscillator, this gives

$$E = \left(v + \frac{1}{2}\right) h\nu \quad \text{where } v = 0, 1, 2, \ldots \tag{8.16}$$

In order for a bond to increase in vibrational energy from v to $v + 1$, it must absorb energy of a particular frequency. From our earlier discussion of the harmonic oscillator in Chapter 1, we know that this frequency will be given by

$$\nu = \frac{1}{2\pi}\sqrt{\frac{k}{\mu}} \tag{8.17}$$

Combining Eqs. (8.16) and (8.17) gives the overall quantum mechanical vibrational energy level result:

$$E = \left(v + \frac{1}{2}\right) \hbar\sqrt{\frac{k}{\mu}} \tag{8.18}$$

In Eqs. (8.17) and (8.18), we employ the **reduced mass** of the system, μ, which takes into account contributions from each atom. It is given by

$$\mu = \frac{m_1 m_2}{m_1 + m_2} \tag{8.19}$$

Strictly we should be speaking of the effective mass here, but for diatomic molecules the reduced mass and effective mass are the same. This is not the case not for polyatomic molecules.

Returning to Eq. (8.16) we should notice that the lowest vibrational energy (when $v = 0$) will be given by

$$E = \frac{1}{2}h\nu \tag{8.20}$$

This implies that molecules will always have some vibrational energy (known as the **zero-point energy**) even at absolute zero. This is a key difference from the behavior of classical springs and cannot be accounted for without involving quantum mechanics. Further application of the Schrödinger equation produces the **specific selection rule** for transitions in vibrational energy levels:

$$\Delta v = \pm 1 \tag{8.21}$$

Taking this into account, we can show that the infrared absorption for a diatomic molecule should consist of a single line. For the $v = 0 \rightarrow v = 1$ transition we would have

$$\Delta E = E_{v=1} - E_{v=0} = \left(v + 1 + \frac{1}{2}\right)h\nu - \left(v + \frac{1}{2}\right)h\nu = h\nu = \frac{h}{2\pi}\sqrt{\frac{k}{\mu}}$$

The observed spectrum will be a single line whose frequency can be used to determine the force constant:

$$\nu_{\text{obs}} = \frac{\Delta E}{h} = \frac{1}{2\pi}\sqrt{\frac{k}{\mu}} \tag{8.22}$$

Of course, this relies on the harmonic approximation. The breakdown of this causes transitions to lie at slightly different frequencies, yielding multiple lines. In addition to the specific selection rule, a **gross selection rule** also applies to transitions in the vibrational energy levels — the molecule must possess an electric dipole moment. For this reason only heteronuclear diatomic molecules are infrared active; homonuclear diatomic molecules are generally infrared inactive (some weak transitions can occur in nanomaterials as a consequence of weak van der Waals attractions).

In a spectroscopic study, we investigate **fundamental transitions** that result from absorption of infrared radiation by the $v = 0$ state. This is

Table 8.2 Physical Data for Selected Covalent Bonds

Bond	Vibration/ cm^{-1}	Force Constant/Nm^{-1}	Internuclear Distance/nm
H–F	4138.5	966	0.0927
H–Cl	2990.6	516	0.1274
C=O	2169.7	1902	0.1131
N=O	1904.0	1595	0.1151

because at room temperature most of the molecules are in the lowest allowed energy state. We can verify this for hydrogen chloride by first calculating the reduced mass of the system:

$$\mu = \frac{1 \times 35.45}{1 + 35.45} \times 1.66 \times 10^{-27} = 1.61 \times 10^{-27}\,\text{kg}$$

The factor 1.66×10^{-27} is required to convert the mass from atomic mass units to kg. Then, using Eq. (8.17) and the data in Table 8.2, we can predict the resonant frequency for the H–Cl bond[a]:

$$\nu = \frac{1}{2\pi}\sqrt{\frac{516}{1.61 \times 10^{-27}}} \approx 9 \times 10^{13}\,\text{Hz}(=90\,\text{THz})$$

If we now use Eq. (8.4) for the $v = 0 \rightarrow v = 1$ transition at room temperature we would obtain the population of the $v = 1$ state:

$$\frac{N_{v=1}}{N_{v=0}} = \exp\left[\frac{-(6.626 \times 10^{-34}\,\text{J}\cdot\text{s})(9 \times 10^{13}\,\text{s}^{-1})}{(1.38 \times 10^{-23}\,\text{J}\cdot\text{K}^{-1})(298\,\text{K})}\right] = 0.000000503$$

That is, virtually 100% of the molecules are present in the $v = 0$ state. Even at 1000 K, 98% of the HCl molecules would be present in the ground vibrational state.

The dependence of resonant frequency on the reduced mass means that the presence of isotopes should produce spectroscopically distinct signals (the force constant is dependent on electron motion and will be the same for different isotopes). We can account for this relatively easily by considering

---

[a]It is customary in infrared spectroscopy to quote wavenumbers rather than frequencies. Hence, 90 THz becomes 3010 cm^{-1}. Most covalent bonds resonate between 4000 and 200 cm^{-1} which is the mid-infrared region.

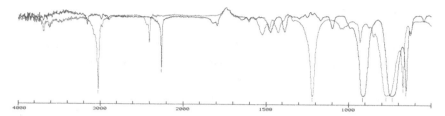

Figure 8.11 Isotope effects in infrared spectroscopy. Chloroform and deuterochloroform have distinctive infrared spectra due to the effect of isotopic substitution.

the ratio of the reduced masses:

$$\frac{\nu_i}{\nu} = \sqrt{\frac{\mu}{\mu_i}} \tag{8.23}$$

where the subscript i refers to the isotope under study. For example, consider the compound trichloromethane, $CHCl_3$, and its deuterated analogue, $CDCl_3$. Infrared spectroscopy of the trichloromethane will reveal a band at $3020 \, cm^{-1}$ assigned to the C$-$H bond. We can then use Eq. (8.23) to predict the position of the C$-$D band:

$$\nu_i = \nu\sqrt{\frac{\mu}{\mu_i}} = 3020 \times \sqrt{\frac{1.53 \times 10^{-27}}{2.85 \times 10^{-27}}} = 2212 \, cm^{-1}$$

Therefore, the spectral line is shifted showing that the separation of the vibrational energy levels is affected by the composition of the nucleus (Figure 8.11). Isotope substitution experiments can prove incredibly useful when trying to assign infrared bands to specific bonds.

One of the great applications of infrared spectroscopy of isotopes is in astrochemistry as it can (*inter alia*) be used to measure the stellar $^{12}C/^{13}C$ ratio. This provides an indication of the progression of the CNO cycle in stars which is known to reduce the $^{12}C/^{13}C$ ratio. The spectroscopic fingerprints of ^{12}CO and ^{13}CO are sufficiently distinct due to the isotopic shift in frequencies to allow estimation of the $^{12}C/^{13}C$ ratio. Similar observations can be made using the ratios of $^{16}O/^{17}O$ and $^{16}O/^{18}O$.

8.5 Mass Spectrometry

Key Point: Mass spectrometry separates atomic and molecular ions according to their mass-to-charge ratio. It is a key technique in isotopic analysis.

Mass spectrometry is a destructive analytical technique used to identify the isotopes present in a compound or to provide structural information for more complex molecules. The technique is based on the creation of ions which are separated in a magnetic field according to their **mass-to-charge ratio**, m/z. Kenneth Bainbridge (1904–1996) is credited with developing the first accurate mass spectrometer, building on the work of Francis Aston (1877–1945) who first used to technique to identify isotopes of non-radioactive elements.

The first stages of mass spectrometry are **vaporization** and **ionization** which leads to the formation of gaseous ions, typically through electron impact:

$$M \rightarrow M(g) + e^- \rightarrow M^+ + 2e^-$$

The next stage is **acceleration**, where ions of charge q are accelerated by a small potential difference, V, increasing their kinetic energy by an amount equal to the product qV:

$$KE = qV = \frac{1}{2}mv^2 \tag{8.24}$$

The penultimate stage is **deflection** where ions are separated according to their m/z ratio. When the ions enter the magnetic sector they will experience a magnetic force, F_M, given by the relationship:

$$F_M = qvB \tag{8.25}$$

where B is the magnetic field measured in tesla, T.[b] The magnetic force constrains the ion to a circular path giving it centripetal force, F_c:

$$F_c = \frac{mv^2}{r} \tag{8.26}$$

For an ion to move in a circular path the two forces must be equal, i.e., Eq. (8.25) = Eq. (8.26), and by rearranging the result and incorporating the square of Eq. (8.24), we arrive at an expression for the mass-to-charge ratio:

$$qvB = \frac{mv^2}{r} \Rightarrow v = \frac{qBr}{m} \quad \text{and} \quad qV = \frac{mv^2}{2}$$

$$\therefore \frac{m}{q} = \frac{B^2r^2}{2V} \quad \text{or} \quad \frac{m}{z} = \frac{B^2r^2e}{2V} \tag{8.27}$$

The ions can be separated by varying the magnetic field strength, before finally reaching the **detection** stage. This usually involves measuring the electrical charge of the ions.

---

[b]The tesla is the SI unit of magnetic field strength. As many magnetic fields are only a fraction of a tesla, often an alternative unit, gauss, G, is used where 1 T = 10^4 G.

Mass spectra show the abundance of a particular ion with at a measured m/z. At their simplest, mass spectra will show the isotopes present in a particular compound. If the instrument is correctly calibrated, the mass of the isotope will be accurate to at least four decimal places and be reported in **atomic mass units** (amu). For elemental analysis, inductively coupled plasma mass spectrometry (ICP-MS) is typically used. This is an incredibly sensitive technique which has a limit of detection of *ca.* 0.02 parts per billion.

The mass spectrometry of compounds, particularly organic molecules, is more complex due to the formation of **molecular fragment ions**. Take the analysis of propane by mass spectrometry (Figure 8.12). At the ionization stage, a propane radical will be formed which will create a peak at $m/z = 44$; this is known as the **molecular ion peak**. A small peak to the right of this is known as the [M+1] peak and is due to the presence of carbon-13. The largest peak (at $m/z = 29$) is known as the **base peak** and is due to the formation of an ethyl fragment ion:

$$CH_3CH_2CH_3 + e^- \rightarrow CH_3CH_2CH_3^+ + 2e^- \rightarrow CH_3CH_2^+ + CH_3^{\cdot}$$

The breaking of the C−C bond also forms a methyl radical with $m/z = 15$. The various other peaks are due to further fragmentation of the molecule.

The isotopic composition of a molecule can be useful when determining the structure of compounds. For example, a sample of pyridine (C_5H_5N; $m/z = 79$) will contain carbon-13, producing a [M+1] peak at $m/z = 80$.

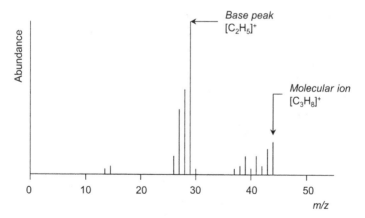

Figure 8.12 Simplified mass spectra of propane. Although the molecular ion and base peaks provide useful information and points of reference, the fragmentation pattern is of most use as it is unique for propane.

However, pyridine will also contain nitrogen-15 which contributes to the [M+1] peak. The resultant intensity of the [M+1] peak can be determined by taking into account the natural abundance of each isotope and the proportions of each element:

$$\text{Intensity}\,[M + 1] = 1.11C + 0.36N \tag{8.28}$$

For pyridine this would be

$$\text{Intensity}\,[M + 1] = (1.11 \times 5) + (0.36 \times 1) = 5.91\%$$

The relative intensity of this peak to the molecular ion peak can then be used to determine the empirical formula of the molecule. A similar argument can be made for the [M+2] peak which has contributions from sulfur, chlorine, and bromine.

One of the most widely known uses of mass spectrometry is **radiocarbon dating** which enables estimation of the age of archeological specimens within the past 50000 years. The technique is based on the measurement of carbon-14, a radioactive isotope of carbon produced in the upper atmosphere from neutron bombardment of nitrogen:

$$^{14}_{7}\text{N} + ^{1}_{0}n \rightarrow ^{14}_{6}\text{C} + ^{0}_{1}p$$

The carbon-14 initially combines with oxygen to form ^{14}CO which then undergoes further reactions with hydroxyl radicals to form carbon dioxide:

$$^{14}\text{CO} + \text{OH}^{\cdot} \rightarrow ^{14}\text{CO}_2 + \text{H}^{\cdot}$$

After equilibrating with non-radioactive carbon dioxide, the $^{14}\text{CO}_2$ enters the biosphere through photosynthesis and the hydrosphere by dissolving in the oceans with an overall abundance ratio of

$$\frac{^{14}_{6}\text{C}}{^{12}_{6}\text{C}} \approx 1.3 \times 10^{-13}$$

During an organism's life cycle, the ratio of carbon-14 is maintained at a constant level as it is in equilibrium with the atmosphere, but on death the carbon-14 levels decrease due to radioactive decay:

$$^{14}_{6}\text{C} \rightarrow ^{14}_{7}\text{N} + ^{0}_{1}e + \bar{\nu}_e$$

Given $t_{1/2} = 5730$ years ($\lambda = 1.21 \times 10^{-4}$ year^{-1}) for carbon-14, the ratio of $^{14}\text{C}/^{12}\text{C}$ will decrease by an amount given by

$$N = 1.3 \times 10^{-13}e^{-1.21 \times 10^{-4}t} \tag{8.29}$$

where N is the ratio at the time of measurement and t is the age of the sample in years. This is the basis of age determination by radiocarbon dating, first reported by Willard Libby (1908–1980) and for which he was awarded the Nobel Prize for Chemistry in 1960.

The main strategy for determining the $^{14}C/^{12}C$ ratio is *accelerator mass spectrometry* (AMS) which employs high kinetic energies and a negative ion source to separate isobars like ^{14}C and ^{14}N. Prior to analysis, the sample must be treated to remove any contaminants, oxidized to carbon dioxide and then converted to graphite *via* the Bosch reaction:

$$\text{Oxidation}: \ C\,(\text{sample}) + 2CuO \rightarrow 2Cu + CO_2$$
$$\text{Bosch Reaction}: \ CO_2 + 2H_2 \rightarrow C + H_2O$$

The $^{14}C/^{12}C$ ratio is then determined and the age of the sample calculated by Eq. (8.29) to give the conventional radiocarbon age before present. This method involves comparison of the sample to an oxalic acid standard prepared in 1950, which predates large-scale testing of nuclear weapons (which altered the $^{14}C/^{12}C$ ratio). Often corrections are made for isotopic fractionation, the mass-dependent fluctuation of carbon isotope ratios in particular environments. This can be easily achieved by measuring the $^{13}C/^{12}C$ ratio, known as $\delta^{13}C$, and subtracting the result from the $^{14}C/^{12}C$ ratio.

Measurement of other isotope ratios can be used for dating. For geological samples, the uranium–lead method is often used, which was pioneered by Clair Patterson (1922–1955) who famously used it to estimate the age of the Earth. The method is based on the measurement of uranium and lead in a sample of zircon — a mineral which incorporates uranium during its formation, but expels lead. As the uranium decays to lead, the former isotope will be released and the ratio of the two can be used to determine the age of the sample.[c] Other isotope combinations, such as rubidium–strontium and potassium–argon have been used to date meteorites and samples from lunar landings.

Chapter Summary

- Nuclear properties can have a significant impact on the behavior of a molecule, particularly with regard to how it behaves when exposed to electromagnetic radiation.

---

[c]There are two decay chains involved here: $^{238}U \rightarrow {}^{206}Pb$ and $^{235}U \rightarrow {}^{207}Pb$.

- Magnetic resonance techniques employ radiofrequency or microwave radiation to investigate the structure of molecules. In NMR, nuclei with odd-numbered spin exert a magnetic moment, which is dependent on its local environment. In EPR, a single unpaired electron moves between excited and ground states, producing a spectroscopic signature unique to its environment.
- Certain nuclei interact with gamma ray photons to produce a unique Mössbauer spectrum, which can be used to determine the oxidation state of an element. This is particularly, useful in remote sensing, such as on the Mars rover.
- Isotopic substitution alters the behavior of covalent bonds when exposed to infrared radiation. This infrared spectroscopic data can be used to determine isotope ratios in stellar phenomena.
- Absolute proportion of isotopes can be determined accurately by mass spectrometry. This forms the basis of radiocarbon dating and also structural determinations.

Review Questions

(1) Express the wavelength 670 nm in terms of frequency and wavenumber.
(2) Approximate the lifetime of the state which has an uncertainty in its frequency of 10^7 Hz.
(3) The wavenumber associated with the fundamental vibration of the bond in a Cl_2 molecule is $565\,cm^{-1}$. Calculate the force constant of the bond.
(4) The frequency of vibration of $^1H^{35}Cl$ is $2990.6\,cm^{-1}$. Without calculating the force constant, evaluate the frequency of vibration for $^1H^{37}Cl$, $^2D^{35}Cl$, and $^2D^{37}Cl$.
(5) Calculate the energy separation between the spin states of an electron exposed to a magnetic field of 0.3 T.
(6) Sketch the ^1H-NMR spectrum for CH_3CHBr_2 indicating the origin of each of the peaks.
(7) What effect does isotopic substitution of deuterium for hydrogen have on the appearance of ^1H and ^{13}C NMR spectra?
(8) Predict the appearance of the 1H-NMR spectrum for deuterated acetone, CD_3COCD_3 assuming no coupling between the CD_3 and CD_2H groups.

(9) A free Mossbauer nucleus of mass 1.67×10^{-25} kg emits a gamma ray photon with a wavelength of 0.1 nm. Calculate the recoil velocity of the nucleus.

(10) In an experiment, the mass spectrum of two isomeric ketones, 2-hexanone and 4-methyl-2-pentanone, were recorded. Predict (a) the m/z for the molecular ion peak; (b) the structure of the base peak at m/z 58; (c) the identity of the peaks at m/z 15, 18, and 85.

Chapter 9

Applications of Nuclear Chemistry

"This work on deuterium is only the beginning of a very interesting scientific development."

H. Urey

Nuclear chemistry is a large field, which incorporates many aspects of biology, chemistry, and physics. Two main branches will be discussed here: radiochemistry and radiation chemistry. On completion of this chapter and the associated questions, you should:

- Have a quantitative understanding of equilibrium isotope effects and kinetic isotope effects.
- Understand the role of isotopes in practical chemistry, particularly with regard to isotope dilution techniques and determination of physicochemical constants.
- Be able to apply key concepts in radiation chemistry to chemical and biological systems.

9.1 Introduction

Key Point: Nuclear chemistry is broadly divided into radiochemistry and radiation chemistry.

Nuclear chemistry is a very broad discipline; it includes the behavior of all isotopes, both stable and radioactive. A subdivision of nuclear chemistry, referred to as **radiochemistry**, deals with the chemistry of radioactive isotopes and how ionizing radiation can be used to study the properties of stable isotopes. This could include labeling organic compounds with radioactive tracers or the use of radioisotopes in medical imaging. In the broader sense, radiochemistry also includes areas such as the environmental effects of radioactive waste disposal, regulation of radioactive materials, and safeguards for the non-proliferation of nuclear materials. On the other hand, *radiation chemistry* investigates the effect of radiation on chemical behavior; no radioactive material needs to be involved in the process under study.

The development of nuclear chemistry can be traced back to the work of American physical chemist Harold Urey (1893–1981), who together with Ferdinand Brickwedde (1903–1989) discovered deuterium in 1932. The role of deuterium in the nuclear revolution has been well documented and its use in practical chemistry made techniques such as NMR possible. The majority of world's supply of deuterium comes from *heavy water* (deuterium oxide, ^2H_2O often written as D_2O). Early methods, such as those employed at the Norsk Hydro plant in Norway in the 1940s, used electrolysis to enrich natural water with deuterium oxide. An alternative process developed in the United States is the *Girdler sulfide process* (Figure 9.1), which is based on the isotope exchange equilibrium:

$$H_2O + HDS \rightleftharpoons HDO + H_2S$$

The process is exothermic: at $30°C$ $K = 2.33$ while at $130°C$ $K = 1.82$. Therefore, when hydrogen sulfide gas is circulated in a closed loop between hot and cold areas, deuterium passes from the hydrogen sulfide to the water. The process is repeated until the water is enriched to about 20%; further enrichment requires distillation or electrolysis.

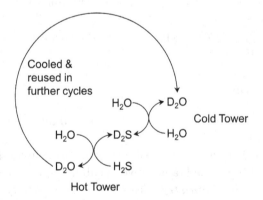

Figure 9.1 The Girdler sulfide process. The deuterium content of water is enriched by successive phase equilibrium separations with hydrogen sulfide.

9.2 Isotopes Effects

Key Point: The presence of isotopes can alter the physicochemical properties of a molecule, such as the position of equilibrium, and the rate of reaction.

It is often stated that isotopes of the same element have the same chemical properties. At a cursory glance, this may appear to be true, but the impact of the nucleus on surrounding electrons can produce significant effects, particularly for lighter elements where the valance electrons are closer to the nucleus. Differences in chemical or physical behavior brought about by isotopic substitution in a molecule are referred to as *isotope effects* — for example, in the previous chapter, we saw how the isotopic composition of chloroform altered its infrared spectrum. Molecules that differ only in their isotopic composition are known as *isotopologues* — for example, H_2O, DHO, and D_2O. If the molecule is an isomer with the same number of substituted isotopes, but in a different arrangement, it is referred to as an *isotopomer* — for example, $CH_2DCH=O$ and $CH_3CH=O$.

9.2.1 Equilibrium isotope effects

Key Point: At equilibrium, the lighter isotope tends to migrate to the compound of lower bond energy, causing isotope fractionation.

The first most noticeable equilibrium isotope effect is that found in the autoionization of water, $2H_2O \rightleftharpoons H_3O^+ + OH^-$, which has an equilibrium constant of 1.00×10^{-14} at 25°C. If we consider the same process for deuterium oxide, $2D_2O \rightleftharpoons D_3O^+ + OD^-$, a significantly lower value is obtained (1.38×10^{-15}), which implies that the deuterium ion does not dissociate as readily as hydrogen. This effect arises from deuterium's lower zero-point energy, which increases the $O-D$ bond dissociation enthalpy $(H_2O, 458.9\,kJmol^{-1}$ versus $D_2O, 492.2\,kJmol^{-1})$. A similar effect can be observed when weak acids are dissolved in D_2O, where the isotope effect reduces the observed acidity of the acid.

In the more general case, we can qualify the isotope effect on the equilibrium constant by considering two equilibria. The first involves the lighter isotope of element X, denoted as X':

$$AX' + BY \rightleftharpoons BX' + AY \quad K_1 = \frac{[BX'][AY]}{[AX'][BY]} \tag{9.1}$$

In the second equilibria, a heavier isotope of element X has been substituted:

$$AX + BY \rightleftharpoons BX + AY \quad K_2 = \frac{[BX][AY]}{[AX][BY]} \tag{9.2}$$

If we combine Eqs. (9.1) and (9.2) to cancel out the common terms, [AY] and [BY], we will obtain the equilibrium constant for the isotope exchange process:

$$AX' + BX \rightleftharpoons AX + BX' \quad K = \frac{K_1}{K_2} = \frac{[BX'][AX]}{[AX'][BX]} \tag{9.3}$$

The deviation from unity gives an indication of the size of the isotope effect with the lighter isotope having a tendency to concentrate in the compound with smaller bond energy. The fact that the equilibrium constant for isotope exchange reactions is not unity can be used for the separation of isotopes, as we saw with the Girdler sulfide process. A further example is the exchange reaction between nitrogen(II) oxide and nitric acid,

$$^{15}NO(g) + H^{14}NO_3(aq) \rightleftharpoons {}^{14}NO(g) + H^{15}NO_3(aq)$$

which has $K = 1.05$ at $25°C$. In this example, the gaseous nitrogen(II) oxide will bubble out of the nitric acid solution shifting the position of equilibrium the right-hand side, to enriching the nitrogen-15 content. This process is often termed *isotope fractionation,* in which the equilibrium constant is replaced by the *fractionation factor*, α:

$$\alpha = \frac{R_{AX}}{R_{BX}} \tag{9.4}$$

where R_{AX} and R_{BX} are the isotope ratios for the general process shown in Eq. (9.3). From this, we see that if one atom from each molecule participates in the exchange, $K = \alpha$. Conversely, the exchange reaction:

$$C^{16}O + {}^{16}O^{18}O \rightleftharpoons C^{18}O + {}^{16}O_2 \tag{9.5}$$

will have $K = 1/2\alpha$ because there are two oxygen atoms in molecular oxygen so the oxygen isotope ratio is less affected by reaction.

We can use statistical thermodynamics to evaluate the equilibrium constant (and fractionation factor) for isotope exchange processes. To demonstrate the theory, we first break down the equilibrium constant for Eq. (9.5) into three individual equilibrium constants:

$$K = K_{\text{trans}} \cdot K_{\text{rot}} \cdot K_{\text{vib}} \tag{9.6}$$

where K_{trans}, K_{rot}, and K_{vib} are the equilibrium constants arising from translational, rotational, and vibrational partition functions. These can be evaluated to a good approximation from Eqs. (9.7) to (9.9):

$$K_{\text{trans}} = \left(\frac{M_{C^{18}O} \cdot M_{^{16}O_2}}{M_{C^{16}O} \cdot M_{^{16}O^{18}O}} \right)^{3/2} \tag{9.7}$$

$$K_{\text{rot}} = \frac{I_{C^{18}O} \cdot I_{^{16}O_2}}{2I_{C^{16}O} \cdot I_{^{16}O^{18}O}} \tag{9.8}$$

$$K_{\text{vib}} \approx 1 + \frac{h}{2kT}[(\nu_{C^{16}O} - \nu_{C^{18}O}) - (\nu_{^{16}O_2} - \nu_{^{16}O^{18}O})] \tag{9.9}$$

where M is the molecular mass, I is the moment of inertia (the product of the reduced mass and the bond distance squared, μd^2) and ν is the vibrational frequency of the specified bond. Thus, the translational equilibrium constant will be given by

$$K_{\text{trans}} = \left(\frac{32 \times 30}{34 \times 28} \right)^{3/2} = 1.0126$$

Evaluation of the rotational equilibrium constant from Eq. (9.8) can be simplified, as bond lengths are virtually independent of the isotope involved; therefore, K_{rot} can be determined directly from the reduced masses:

$$K_{\text{rot}} = \frac{1}{2} \left(\frac{8 \times 7.2}{8.5 \times 6.8} \right) = 0.4982$$

Finally, the vibrational equilibrium constant is evaluated from Eq. (9.9) by recognizing that the vibrational frequencies are related to the reduced masses by the harmonic oscillator (*c.f.* Chapter 8):

$$\nu_{C^{18}O} = \nu_{C^{16}O} \sqrt{\frac{\mu_{C^{18}O}}{\mu_{C^{16}O}}} = 0.976\nu_{C^{16}O} \quad \text{and}$$

$$\nu_{^{16}O^{18}O} = \nu_{^{16}O_2} \sqrt{\frac{\mu_{^{16}O^{18}O}}{\mu_{^{16}O_2}}} = 0.9718\nu_{^{16}O_2}$$

Using the vibrational frequencies (obtained from infrared spectroscopy) for CO and O_2, 6.50×10^{13} s^{-1} and 4.74×10^{13} s^{-1}, respectively, we obtain the vibrational equilibrium constant (at 300 K):

$$K_{\text{vib}} = 1 + \frac{h}{2kT}\{[6.50 \times 10^{13}(1 - 0.976)]$$

$$- [4.74 \times 10^{13}(1 - 0.9718)]\} = 1.0199$$

The product of these values gives $K = 0.5145$ and $\alpha = 1.029$. On this basis, we would expect to see enrichment of oxygen-18 in the carbon dioxide, which could be separated from the other components by careful fractional distillation. The temperature dependence of Eq. (9.9) means that our approximations only hold at relatively low temperatures.

9.2.2 Kinetic isotope effects

Key Point: The difference in zero-point energy between two isotopes produces a difference in activation energy; this changes the rate of reaction in what is known as a kinetic isotope effect.

The kinetic isotope effect describes how the rate of reaction changes with isotopic substitution and has proven to be a valuable tool for investigating reaction mechanisms. The effect depends on how the change in atomic mass affects the zero-point energy: thus, when hydrogen is replaced by deuterium, the mass doubles and the rate constant for the reaction changes.

Consider the conversion of 1-bromopropane to propene using sodium ethoxide (Figure 9.2). If this reaction were to proceed in a single step, the rate-determining step would involve the removal of a hydrogen atom from the second carbon. If we were to replace the labile hydrogen atoms in the alkyl halide with deuterium and compare the rate constants, any observed change in the rate constant would verify the participation of a methylene hydrogen atom. Following such an experiment, the ratio of the two rate constants, k_H/k_D, was found to be 6.4. This is consistent with a ***primary kinetic isotope effect*** — the rate constant changes due to isotopic substitution of a bond involved in the rate-determining step.

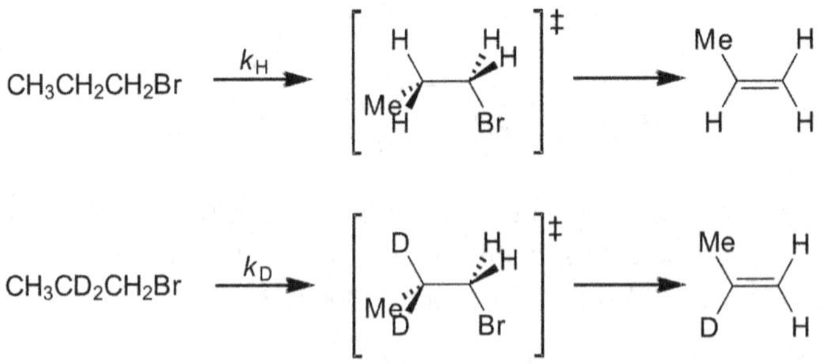

Figure 9.2 Kinetic isotope effect. In the E2 elimination of 1-bromopropane, the rate-determining step is dependent on the bond energy of the C–H and C–D bonds, which is in turn dependent on the zero-point energy of the system.

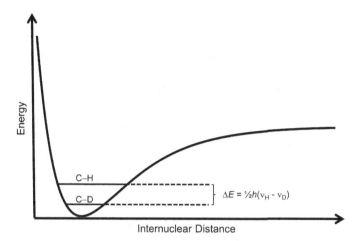

Figure 9.3 Morse potential and kinetic isotope effect. The difference in zero-point energy between the C–H and C–D bonds can be related to their vibrational frequencies. This allows us to approximate the ratio of the rate constants.

A physicochemical explanation for this can be obtained by plotting the Morse potential for the C–H/C–D system (Figure 9.3). We see that the zero-point energy for the C–H bond is higher than that of the C–D bond, which means that the activation energy for the reaction involving the C–D bond will higher. We can use the fact that kinetic isotope effects are dominated by zero-point energy differences to approximate the change in the rate constant. The difference in activation energy can be determined by recalling that the zero-point energy is given by

$$E = \frac{1}{2}h\upsilon \tag{9.10}$$

We have already seen in Chapter 8 that a C–H bond vibrates at a frequency of $3020\,\text{cm}^{-1}$ ($9.03 \times 10^{13}\,\text{Hz}$) and that the expected frequency of the C–D bond is $2212\,\text{cm}^{-1}$ ($6.61 \times 10^{13}\,\text{Hz}$). The difference between vibrational energy levels (and activation energy) is therefore,

$$\Delta E_a = -\frac{1}{2}h(\upsilon_H - \upsilon_D) \tag{9.11}$$

We can relate this to the rate constant, k, for each reaction by way of the Arrhenius equation:

$$k_H = Ae^{-E_a/kT} \quad \text{and} \quad k_D = Ae^{-E_a/kT} \tag{9.12}$$

If we assume that the pre-exponential factor is independent of any isotope effect, the ratio of the rate constants will be related to the difference in activation energy[a]:

$$\frac{k_H}{k_D} = \exp\left(-\frac{\Delta E_a}{kT}\right) = \left[\frac{1/2h(\nu_H - \nu_D)}{k_B T}\right] \qquad (9.13)$$

which at $298\,K$ returns $k_H/k_D \approx 7$, which is typical for a deuterium kinetic isotope effect. Due to the dependence on zero-point energy, the kinetic isotope effect will increase the lower the temperature. A more rigorous treatment of the kinetic isotope effect involves using partition functions, which take into account the bending vibrations and other changes in the geometry of the molecule. These calculations are now easily performed using commercially available computational chemistry software, such as Gaussian.

Secondary kinetic isotope effects are observed when the rate constant is affected by isotopic substitution of bonds not involved in the rate-determining step. These usually fall into two categories:

(1) Normal $2°$-KIE: $k_H/k_D > 1$ when bond changes from $sp^3 \to sp^2$.
(2) Inverse $2°$-KIE: $k_H/k_D < 1$ when bond changes from $sp^2 \to sp^3$.

The presence of an electronegative atom can also attribute to the kinetic isotope effect. For example, the dissociation of ethanoic acid

$$CH_3COOH \rightleftharpoons CH_3COO^- + H^+$$
$$CD_3COOH \rightleftharpoons CD_3COO^- + H^+$$

has $k_H/k_D = 1.06$ implying that hydrogen is more electronegative than deuterium. This is known as an inductive kinetic isotope effect.

9.3 Stable Isotope Tracers

Key Point: Stable isotopes are commonly incorporated into molecules to act as internal standards for analytical procedures. A common use is in isotope dilution analysis.

Aside from their use in determining reaction mechanisms, stable isotopes can be employed to follow the movement of a species or functional group in a chemical reaction. This is possible for the vast majority of elements as kinetic isotope effects are only significant for very light elements like

---

[a]A similar approach common in physical chemistry texts is to use the change in Gibbs energy and the Eyring equation.

hydrogen. Tracer studies with stable isotopes usually employ ^{2}H, ^{13}C, ^{15}N, and ^{18}O, which can be easily identified by mass spectrometry or NMR.

The nomenclature for tracers follows that commonly used for compounds. For example, carbon-13 labeling of acetic acid could be achieved in three ways:

(1) Labeling at C_1 gives [1-^{13}C]acetic acid (^{12}CH$_3^{13}$COOH)
(2) Labeling at C_2 gives [2-^{13}C]acetic acid (^{13}CH$_3^{12}$COOH)
(3) Uniform labeling gives [U-^{13}C]acetic acid (^{13}CH$_3^{13}$COOH)

The term **uniform labeling** means that the isotope is uniformly distributed throughout the molecule. For larger molecules the prefixes r, G and N are often used to indicate random, general, and nominal labeling, which all mean the distribution of the label is random. In certain compounds with their own established nomenclature, other locants may be used.

Stable isotope tracers are synthesized to incorporate a specific amount of the isotope. For example, [^{13}C]-glucose is synthesized in a series of reactions starting from carbon-13 enriched carbon dioxide:

$$^{13}CO_2 + H_2O \rightarrow {}^{13}CH_4 + H_2 \quad \text{and} \quad {}^{13}CH_4 + NH_3 \rightarrow H^{13}CN + 3H_2$$

The hydrogen cyanide is reacted with D-arabinose, which produces a mixture of D-[1-^{13}C]-gluconitrile and D-[1-^{13}C]-mannononitrile, which are then reduced and hydrolyzed to D-[1-^{13}C]-glucose and D-[1-^{13}C]-mannose. The two-labeled sugars can then be resolved using preparative chromatography. Increasingly, biotechnology is being used to label compounds, especially when the stereochemistry of the labeled group needs to be conserved.

9.3.1 Stable isotope dilution analysis

Key Point: A known quantity of stable isotope is combined with a sample; the ratio of isotopic masses can be used to calculate the quantity of analyte in the sample.

Stable isotopes are often used to improve the sensitivity of analytical methods, particularly those involving mass spectrometry, in what is essentially a modification of the standard addition method. The principle of stable isotope dilution is simple: a sample is mixed with an internal standard, which is labeled with an appropriate stable isotope. Since both the target analyte and internal standard are chemically identical, they will co-separate by whatever means is employed, but will produce two different peaks during mass spectrometry. As the sample and standard are being subjected to exactly the same process, any losses during the workup are

incurred by both and the relative loss will be effectively zero. When the
ratio of the isotopes is measured, the quantity of the unknown can be
calculated:

$$C_x = \left(\frac{C_s m_s}{m_x}\right)\left(\frac{A_s - RB_s}{RB_x - A_x}\right) \qquad (9.14)$$

where C_x is the concentration of the unknown, C_s is the concentration of
the internal standard, m_s is the mass of the internal standard, m_x is the
mass of the sample, A_s and B_s are the atom fractions of isotope A or B in
the internal standard, A_x and B_x are the atom fraction of isotope A or B
in the sample, and R is the measured ratio of isotopes.

To illustrate the principle, consider the measurement of vanadium in a
sample of crude oil. Vanadium exists naturally as the vanadium-51 isotope
at an abundance of 99.75% while the vanadium-50 isotope is present at
0.0025%. If we take a 0.401 g sample of crude oil and add 0.419 g of
vanadium standard (2.245 μmol vanadium/g enriched with vanadium-50–
36.09%) and obtain an isotope ratio of 10.545 by mass spectrometry, the
concentration of vanadium in the sample would be

$$C_x = \left(\frac{2.245 \times 0.419}{0.401}\right)\left(\frac{63.91 - 10.545 \times 36.09}{10.454 \times 0.0025 - 99.75}\right) = 7.448\,\mu\text{mol/g}$$

In order for this technique to work, the ratio of isotopes in the internal
standard must be known. Certified reference materials are available for
most routine analyses; for other work, the isotope ratio must first be
determined in the internal standard. Determining the composition of the
internal standard is known as *reverse isotope dilution*. It involves
taking a standard with a natural known isotopic abundance and using
it to determine the composition of the enriched standard using the same
approach as above. Often the two steps are combined in what is known as
double isotope dilution.

9.4 Radioactive Tracers

Key Point: Radioactive tracers provide a sensitive means of determining
key physical constants. They can also be used to establish physiological
measurements in humans or laboratory animals.

The use of radioactive tracers greatly improves the sensitivity of the
experimental technique — typical tracers allow detection of picogram
quantities — and their relative ease of measurement makes their use
appealing. During the preparation of the tracer, a quantity of the tracer's

stable isotope(s) often remain, which are referred to as the **carrier**. If the concentration of the stable isotope(s) is low, it is regarded as being **carrier free**. The **radiochemical purity** of the tracer is the total radioactivity in the tracer, while the **radionuclide purity** is the quantity of the radioactivity due to the stated radionuclide.

When preparing a radioactive tracer, we first consider the **maximum specific activity** of the radionuclide involved. For example, carbon-11 ($t_{1/2} = 20.33\,\text{min}$) will have a maximum specific activity given by

$$\left(\frac{\ln 2}{t_{1/2}}\right) \times 6.02 \times 10^{23} = 3.42 \times 10^{20}\,\text{Bq/mol}$$

which would be usually written as $3.42 \times 10^5\,\text{GBq}/\mu\text{mol}$. If the carbon-11 is then used to label acetic acid at both positions, the maximum specific activity would be

$$\frac{3.42 \times 10^{20} \times 2}{64.036} = 1.07 \times 10^4\,\text{GBq}/\mu\text{mol}$$

Of course, the actual specific activity of the labeled acetic acid will be much lower, probably around 1500–$2500\,\text{GBq}/\mu\text{mol}$. This is because it is impossible to avoid dilution by stable isotopes. These can be incorporated at the time of producing the radionuclide itself, or during the synthesis of the tracer.

The synthesis of radioactive tracers involves much of the same chemistry as for stable isotopes. Labeling with tritium was historically performed by the **Wilzbach method**, in which the compound to be labeled was combined with an excess of tritium gas. The mechanism involves ionization of the target molecule by the beta particles released from the tritium. This method is obviously very inefficient and non-specific and would only be used as a last resort in modern radiochemistry. Modern techniques now involve either catalytic isotope exchange or direct chemical synthesis (Figure 9.4). Catalytic exchange tends to be reserved for labeling with tritium, using catalysts, such as aluminum chloride, phosphoric acid-boron trifluoride, or palladium. Alternatives to the catalytic process could involve simple functional group transformations. For example, if a tritiated aldehyde is required, a commercially available tritiated alcohol could be oxidized by heating for a few minutes with potassium dichromate. Labeling with isotopes of carbon may include classic organic chemistry or enzyme-catalyzed methods and usually start with labeled precursors (e.g., $^{11}CO_2$, $^{14}CH_4$, etc.). Carbon-14 is quite amenable to conventional microscale chemistry and can be incorporated

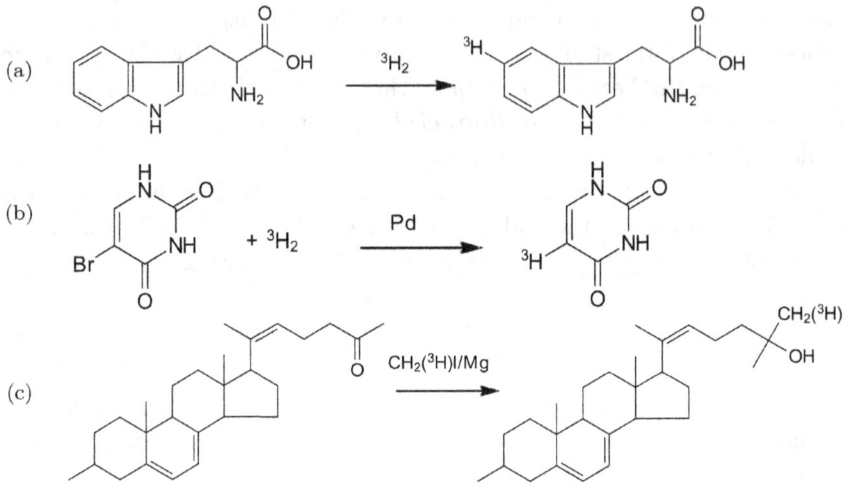

Figure 9.4 Radiolabeling of organic compounds. (a) Wilzbach method; (b) catalyzed isotope exchange; (c) Grignard reaction.

into a range of compounds using standard techniques. The labeling of biological material is more complex as often-lengthy separation procedures are required.

When the tracer has been synthesized, unless it is used immediately, it will undergo a reduction in activity due to radioactive decay and so an adjustment is required. For example, suppose we had a 10 mL supply of tritiated toluene, which was manufactured with a specific activity of 5.03 kBq/g. If we wanted to use 5 mL of the toluene one year later, we would have to adjust the specific activity to take into account the decay of the tritium. Taking $t_{1/2} = 12.26$ years we obtain,

$$A = A_o e^{-\lambda t} = 5.03 \cdot \exp\left(-\frac{\ln 2}{12.26} \times 1\right) = 4.75 \text{ kBq/g}$$

To get the specific activity in a 5 mL aliquot of [³H]-toluene we multiply the result by the density of the toluene,

$$\text{SA} = (4.75 \text{ kBq} \cdot \text{g}^{-1}) \times (0.8669 \text{ g} \cdot \text{mL}^{-1}) = 4.12 \text{ kBq} \cdot \text{mL}^{-1}$$

Calculations of this nature are generally performed automatically using inventory software (e.g., IsoStock), which calculates the decrease in activity since the time of manufacture.

9.4.1 Dilution analysis

Key Point: The ratio of the specific activity of a radioactive tracer to an analyte can be used to precisely determine the quantity of analyte in a complex sample.

The specific activity of a tracer is reduced when we add an excess of carrier. If we take a sample containing an analyte, add a known amount of tracer and perform some form of separation (e.g., chromatography), the specific activity of the separated component is related to the mass of analyte by

$$m = m_s \left(\frac{S_1}{S_2} - 1 \right) \tag{9.15}$$

where m is the mass of analyte, m_s is the mass of the tracer, S_1 is the specific activity of the tracer, and S_2 is the specific activity after separation. For example, consider the determination of the quantity of penicillin in a crude preparation. We could combine the sample with, say, 10 mg of labeled penicillin ($14.9\,\text{kBq/mg}$) and the isolate both using solvent extraction. When the activity of the organic phase was measured, an activity of $1929.5\,\text{kBq/mL}$ was found. If the organic phase was chloroform, which has density $= 1.49\,\text{g/mL}$, the specific activity would be

$$\text{SA} = \frac{1929.5\,\text{kBq} \cdot \text{mL}^{-1}}{1.49\,\text{g} \cdot \text{mL}^{-1}} = 1294.9\,\text{kBq} \cdot \text{g}^{-1}$$

Therefore, by Eq. (9.15), the original amount of penicillin would be

$$m = 10\,\text{mg} \times \left(\frac{14.9\,\text{kBq} \cdot \text{mg}^{-1}}{(1294.9\,\text{kBq} \cdot \text{g}^{-1}) \times (1 \times 10^{-3}\text{g} \cdot \text{mg}^{-1})} - 1 \right) = 105\,\text{mg}$$

This approach is widely used in the analysis of complex samples — for example, the determination of an analyte in serum, which contains hundreds of components.

9.4.2 Radioimmunoassay (RIA)

Key Point: The specific reaction between a labeled antigen and an antibody can be used to determine trace amounts of biologically important molecules such as hormones and regulatory peptides.

A key feature of modern laboratory medicine is the measurement of small quantities of biological molecules of physiological significance.

Until the widespread development of liquid chromatography-mass spectrometry methods, RIA was the only technique capable of detection such small quantities of material. It found particular application in toxicology and hormone detection, where not only the levels of target analyte are incredibly small, but they are also found in a complex sample matrix, necessitating a highly specific assay.

The principle of the technique is based on the specific reaction between an antigen, Ag, and an antibody, Ab:

$$Ag + Ab \rightleftharpoons Ag - Ab \quad K = \frac{[Ag - Ab]}{[Ag][Ab]} \tag{9.16}$$

The equilibrium constant for Eq. (9.16) is typically in the order of 10^{10}, which implies that the formation of the antigen–antibody complex is strongly favored.

In a RIA, a known quantity of antigen is labeled with a radioisotope (usually ^{125}I) and this is mixed with a known quantity of antibody. The sample is then introduced, and the sample antigen displaces a portion of the radiolabeled antigen:

$$Ag + Ag^* + Ab \rightarrow Ag - Ab + Ag^* - Ab + Ag + Ag^* \tag{9.17}$$

The bound antigens are then separated from the free antigen and the radioactivity determined, usually by gamma counting. In practice, a dilution technique is employed, whereby the level of sample antigen is steadily increased. The ratio of the radioactivity of the initial $Ag^* - Ab$ complex (known as B_0) to the radioactivity of the various $Ag/Ag^* - Ab$ complexes formed on addition of sample antigen (B) is related to the amount of sample antigen.

9.4.3 Solubility constants

Key Point: Solubility constants can be readily determined from gravimetric analysis with radioactive tracers.

In 1898, Marie and Pierre Curie first isolated radium from pitchblende and found that it shared many chemical properties with Group 2 elements. With this knowledge, they were able to determine the relative atomic mass of radium through a simple precipitation reaction:

$$RaCl_2 + AgNO_3 \rightarrow 2AgCl + Ra(NO_3)$$

This idea that precipitation reactions could be used to separate radioactive materials was further developed by George Hevesy (1885–1966) and Fritz

Paneth (1887–1958). Hevesy was challenged by Ernst Rutherford to isolate "radium-D" from a sample of lead.[b] After two years of intense experimental work, Hevesy concluded that it was impossible to isolate the radium-D by chemical means. Instead, he concluded that radium-D could be used as an indicator of the amount of lead present. By combining a sample of lead as its nitrate with a small quantity of radium-D and precipitating the lead/radium-D as the chromate, the radioactivity of the precipitate could be related to the total amount of lead. Hevesy and Paneth then applied this methodology to determine the solubility of various lead salts.

The determination of solubility constants from radioactive tracer experiments is relatively straightforward. Suppose we wished to determine the solubility constant for silver bromide. We would first precipitate silver bromide from a solution of its salt using silver nitrate doped with a silver-112 tracer:

$$NaBr + AgNO_3 \rightarrow AgBr + NaNO_3$$

If the measured radioactivity of the precipitate was 17316 cpm and the counting efficiency was 75%, the activity in dpm would be

$$Activity = \frac{17316}{0.75} = 23088 \, dpm$$

If we had 1 mg of the precipitate, the specific activity of the sample would be

$$Specific \, Activity = \frac{23088 \, dpm}{1 \, mg} = 23088 \, dpm \cdot mg^{-1}$$

To establish the solubility equilibrium, we would combine the precipitate with water, then after a period of time remove an aliquot of the supernatant and determine its radioactivity. A typical value would be 3.108 dpm/mL and the ratio of this to the specific activity will give the mass of silver bromine per mL.

$$\frac{3.108 \, dpm/mL}{23088 \, dpm/mg} = 1.35 \times 10^{-4} \, mg/mL$$

This is equivalent to a molar solubility of 7.2×10^{-7} M and a solubility constant of 5.18×10^{-13}. The advantage of this method is that we do not need to know the isotopic composition of the silver nitrate since we are dealing with relative amounts. Similar principles can be applied to determine other equilibrium constants, such as the distribution of a solute between two solvents.

---

[b]Radium-D has since been identified as lead-112.

9.5 Neutron Activation Analysis

Key Point: In neutron activation analysis, stable nuclei are rendered radioactive by neutron bombardment; subsequent decay produces unique spectroscopic signatures, which identify the elements in a sample.

In neutron activation analysis, stable nuclei are rendered radioactive through neutron bombardment; the subsequent decay of the nuclei produces unique gamma ray signatures, which can be related to the quantity of material present. Those elements with larger nuclei are more amenable to neutron activation analysis (larger cross-section for neutron capture). For this reason, the technique is often used for trace determination of rare Earth elements.

Nuclear reactors provide a steady source of thermal neutrons (typical neutron flux, $\phi = 10^{13}$ neutrons cm^{-2}s^{-1}), which can give average detection limits in the region of 10 μg. Alternatively, a bench-top fusor can be used, in which tritium absorbed onto a titanium or zirconium target, is bombarded by deuterons:

$$\,^2_1\text{H} + \,^3_1\text{H} \rightarrow \,^4_2\text{He} + \,^1_0 n$$

Other instruments, particularly portable analyzers, may use isotopes, which release neutrons through spontaneous fission. A common example of this is californium-238; *ca.* 3% of its decay occurs *via* spontaneous fusion, producing 3.8 neutrons per fission ($\phi = 3 \times 10^7$ neutrons cm^{-2}s^{-1}). Similarly, a mixture of an alpha emitter with a light element can produce neutrons through the reaction:

$$\,^9_4\text{Be} + \,^4_2\text{He} \rightarrow \,^{12}_6\text{C} + \,^1_0 n$$

Common alpha emitters used in this reaction is americium, plutonium, or curium.

When the neutron flux interacts with a target nucleus it gains *ca.* 8 MeV of energy by binding the neutron. This excess energy is released by prompt gamma ray emission, for example,

$$\,^{23}_{11}\text{Na} + \,^1_0 n \rightarrow \,^{24}_{11}\text{Na} + \gamma$$

The net rate of formation of the radioactive nuclei from a single isotope is related to the cross-section of the nucleus and the decay of the radioactive nuclei formed:

$$\frac{dN^*}{dt} = N\phi\sigma - \lambda N^* \tag{9.18}$$

where N^* is the number of radioactive nuclei and N is the number of stable target atoms. We can integrate this equation and with some rearrangement obtain an expression for the **saturation factor**, S in terms of experimental rate measurements:

$$R = N\phi\sigma E \left[1 - \exp\left(-\frac{0.693t}{t_{1/2}}\right)\right] = N\phi\sigma ES \tag{9.19}$$

where E is the efficiency of the detector. The saturation factor gives an indication of how long the irradiation should proceed. In general, the irradiation time rarely exceeds five half-lives.

Detection of the gamma ray photons is usually by germanium semiconductor detectors, the output from which is used to produce a gamma ray spectrum (counts versus energy). Each peak on the spectrum corresponds to the gamma ray energy associated with a particular element. As the area under the peak is proportional to the amount of the element present, by comparison with a suitable standard, the amount of a particular element present in a sample can be determined:

$$w_x = \frac{w_s R_x}{R_s} \tag{9.20}$$

where w_x and R_x are the weight and decay rate of the element in a sample, and w_s and R_s are the weight and decay rate of the standard.

9.6 Radiation Chemistry

Key Point: Ionizing radiation interacts with atoms and molecules in three main ways (photoelectric absorption, Compton scattering, and pair production). These usually bring about the ionization of the species.

Radiation chemistry is the study of the chemical changes brought about by the interaction of ionizing radiation with matter. In general, this definition includes photons, which have sufficient energy to ionize molecules (*ca.* 10 eV). The source of ionizing radiation could be a radionuclide or a mechanical source (e.g., linear accelerator), depending on the type of radiation involved. A key measure in radiation chemistry is the **linear energy transfer** (LET), which is the rate at which the radiation loses energy as it moves through the absorbing medium. Low LET species are particles with mass in the region of an electron (e.g., beta particles), while high LET species have greater mass (e.g., alpha particles). Technically, gamma rays should not be described in terms of LET; when they interact with an absorbing medium, they pass through, or interact directly with an atom, ejecting an electron. Often "gamma LET" really implies the LET associated with the secondary electron produced on absorption of a gamma ray photon.

When ionizing radiation interacts with matter, its energy is transferred to the atoms and molecules comprising the absorbing medium. Charged ionizing species interact with electrons through Coulombic interactions. X-rays and gamma rays interact through three main mechanisms:

(1) **Photoelectric absorption**: the energy of the incident radiation is absorbed by an electron, ejecting it and leaving a vacancy in the orbital. The ejected electron brings about ionization of the medium.

(2) **Compton scattering**: the incident radiation loses part of its energy by ejecting an electron and the now scattered incident radiation leaves the atom.

(3) **Pair production**: the incident radiation passes close to the nucleus and becomes converted to positron–electron pair. The positron is eventually destroyed by an annihilation process producing two photons.

Within 10^{-16} s of interaction, the ionizing radiation creates a variety of short-lived activated states; these typically lead to ejection of an electron. In the gas phase, the ejected electron easily overcomes the Coulomb attraction of the nucleus and escapes, completing the ionization process; for condensed phases, the escape of the electron depends on the medium.

9.6.1 Radiolysis

Key Point: Ionization brings about the dissociation of molecules into a variety of ions and free radicals. A key measure of this process is the radiation chemical yield, G, given as mol/J.

When ionizing radiation interacts with molecules, dissociation often follows, which produces a variety of species including cations, anions, free radicals, and free electrons. The source of radiation in the majority of radiolysis work is cobalt-60, an artificial radioisotope of cobalt, which initially undergoes beta decay to an activated nickel nucleus, before releasing two gamma ray photons at 1.17 and 1.33 MeV:

$$^{60}_{27}\text{Co} \rightarrow {}^{60}_{28}\text{Co} + e^- + \gamma$$

In radiolysis experiments, a key task is to determine the mechanism by which ionizing radiation promotes dissociation of atoms. This usually involves kinetic investigations where the rate constants for each step are evaluated to obtain the correct sequence of reactions. Modern-day radiolysis experiments usually involve EPR spectroscopy and/or *pulse radiolysis*. In this latter technique, a short (picosecond) pulse of high-energy electrons is directed at the sample. This causes a transient change in optical absorption, which can be monitored by UV/Vis spectroscopy and the concentration of the species obtained from the Beer–Lambert law:

$$A = \varepsilon c l \tag{9.21}$$

where A is the absorption at a specified wavelength, ε is the molar absorption coefficient, c is the molar concentration and l is the pathlength. Thus, provided that the absorption coefficient is known, the concentration of a species can be determined from absorbance measurements. The absorption coefficient can be determined from Beer's plots using standard linear regression procedures.

Extensive investigations of the radiolysis of water have been made, not only because of its chemical and biological significance, but also due to its role in the nuclear power industry. The radiolysis of pure water is believed to follow a sequence of reactions initiated by the formation of the $H_2O^{\cdot+}$ radical cation and an electron (Figure 9.5). In the absence of any solutes, water is reformed from the various radical and molecular species and no net decomposition of water is observed. This is particularly important when water is used as the coolant in nuclear reactors, as the build-up of an explosive mixture of hydrogen and oxygen may otherwise occur. In the biological context, the radiolysis of water can be particularly harmful as the various species produced can modify biological molecules, a key stage in many human diseases.

In radiation chemistry, the number of product species produced per $100\,\text{eV}$ absorbed is defined as the *radiation chemical yield*, G, which in SI units would be

$$G = \frac{n}{E} \quad (E = 100\,\text{eV} = 1.602 \times 10^{-17}\,\text{J}) \tag{9.22}$$

In the radiolysis of water, the G-value for each of the intermediates can either be determined directly experimentally, or inferred from stoichiometric relationships (Table 9.1). For example, the yield of hydroxonium ions

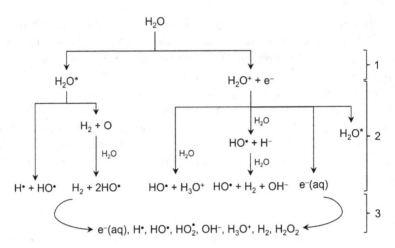

Figure 9.5 Radiolysis of water. After absorbing ionizing radiation, the physical stage (1) occurs with the production of excited water molecules, water cation and an electron. The physicochemical stage (2) involves the formation of a variety of radicals, ions and molecules. Finally, the chemical stage (3) sees the formation of further radicals and molecules such as hydrogen peroxide.

Table 9.1 Radiochemical Yield for Radolysis of Water

Species	e_{aq}^-	$\cdot$OH	H$\cdot$	H$_2$	H$_2$O$_2$	H$_3$O$^+$	$\cdot$HO$_2$
G (μmol/J)	0.28	0.28	0.062	0.047	0.073	0.28	0.0027

Note: $G(\mu\text{mol/J}) = G(\text{mol/100 eV}) \times 0.1036$

must equal the yield of hydrated electrons, which must also equal the yield of hydroxyl radicals:

$$G(e_{aq}^-) = G(\text{H}_3\text{O}^+) = G(\text{OH}^-)$$

In fact, determination of the G-value for the hydrated electron was one of the first triumphs of pulse radiolysis.

9.6.2 Dosimetry

Key Point: The absorbed dose of ionization radiation is measured by dosimetry. The most common chemical dosimeter is the Fricke dosimeter, based on the oxidation of ferrous ions to ferric ions.

A key aspect of any radiolysis experiment is to determine the absorbed dose of radiation. In Chapter 7, we saw that the absorbed gamma ray dose for a human could be evaluated using specific gamma ray constants.

For radiation chemistry experiments, the absorbed dose must be measured by some process, which is similar to the system under study. The most widely used chemical dosimeter is an air-saturated solution of iron(II) sulfate in sulfuric acid, first reported by Hugo Fricke (1892–1972) and subsequently known as the **Fricke dosimeter**. In this system, when the dosimeter is irradiated, ferrous ions are oxidized to ferric ions:

$$e_{aq}^- + H^+ \rightarrow H^{\cdot}$$

$$H^{\cdot} + O_2 \rightarrow {}^{\cdot}HO_2$$

$${}^{\cdot}OH + Fe^{2+} \rightarrow Fe^{3+} + OH^-$$

$${}^{\cdot}HO_2 + Fe^{2+} \rightarrow Fe^{3+} + HO_2^-$$

$$HO_2^- + H^+ \rightarrow H_2O_2$$

$$H_2O_2 + Fe^{2+} \rightarrow Fe^{3+} + OH^- + {}^{\cdot}OH$$

We can evaluate G for ferric ions from those already determined for the radiolysis of water using the stoichiometry of the reaction sequence. Substituting the values from Table 9.1, we obtain

$$G(Fe^{3+}) = 2G(H_2O_2) + 3\left[G(e_{aq}^-) + G(H^{\cdot}) + G({}^{\cdot}HO_2)\right] + G({}^{\cdot}OH)$$

$$= (2 \times 0.073) + [3 \times (0.28 + 0.062 + 0.0027)] + 0.28$$

$$= 1.46\,\mu\text{mol/J} = 1.46 \times 10^{-6}\,\text{mol/J}$$

This allows us to relate the moles of ferric ion produced to the energy of the gamma rays. We can determine the moles of ferric ion from the change in absorbance at λ_{max} 304 nm, which has a molar absorption coefficient of $2.187 \times 10^3\,M^{-1}\,cm^{-1}$.

Recalling the definition of absorbed dose in grey from Chapter 7,

$$D(\text{Gy}) = \frac{E\,(\text{J})}{m\,(\text{kg})} \tag{9.23}$$

we see that this can be related to the G-value, Eq. (9.22), with the units maintaining the dimensions of Gy (J/kg). As the ratio of moles to mass can be shown to be equivalent to the ratio of concentration to density (by cancelling the volume term) we can write

$$D = \frac{n}{G \cdot m} = \frac{c}{G \cdot \rho}$$

Finally, we express c in terms of Eq. (9.21) and substitute into our previous result:

$$D = \frac{A}{G \cdot \rho \cdot \varepsilon \cdot l} \tag{9.24}$$

Evaluating the constants, $G = 1.46 \times 10^{-6} \, \text{mol/J}$, $\varepsilon = 2.187 \times 10^3 \, \text{M}^{-1} \, \text{cm}^{-1}$ and $\rho = 1.024 \, \text{kg/dm}^3$, and taking $l = 1 \, \text{cm}$, Eq. (9.24) can be simplified to

$$D = 306 \cdot A \tag{9.25}$$

where the dose will be in gray. Some versions of Eq. (9.25) have a temperature correction, but with modern thermostatically controlled instruments, this is usually not necessary.

The Fricke dosimeter is linear for a dose up to *ca.* 500 Gy; beyond this a "super Frick dosimeter" can be employed which contains higher concentrations of iron(II) sulfate. A variety of other chemical and instrumental dosimeters is available, depending on the source of radiation involved.

9.6.3 Applications of radiation chemistry

Key Point: Radiation chemistry is used in the synthesis of polymers, in the reduction of pollutants and in the sterilization of medical products.

The fundamental principles of radiation chemistry are widely applied across the sciences. In the early days of radiation chemistry, the use of ionizing radiation to mediate key chemical reactions was intensely investigated. It was found that while relatively simple reactions, such as the Haber–Bosch process had $G < 0.5 \, \mu\text{mol/J}$, other processes involving chain reactions could have very favorable G-values. The Dow Chemical company used radiation-induced addition of hydrogen bromide to ethylene to produce bromoethane, a key feedstock for the chemical industry. In more recent times, the focus of radiation-induced synthetic chemistry has been on polymers. It has been found that when monomer solutions are irradiated by low-LET gamma rays, the radicals formed can initiate the chain polymerization reaction.

An interesting emergent application of radiation chemistry is in environmental protection, where ionizing radiation has been used to degrade pollutants at a faster rate than conventional methods. For example, a variety of metal ions can be recovered from wastewater by the primary products of water radiolysis at around 2.5 kGy. Lead(II) ions, for example, undergo rapid reduction to lead(I) in the presence of hydrated electrons:

$$e_{aq}^- + Pb^{2+} \rightarrow Pb^+$$

The lead(I) can then undergo disproportionation to lead(II) and metallic lead, provided that oxygen-free conditions are maintained (to prevent reoxidation):

$$2Pb^+ \rightarrow Pb^{2+} + Pb$$

Similar approaches can be applied for removal of organic compounds, such as dyestuff or even toxins present in sewage.

A final application of radiation chemistry worth highlighting is its role in food preservation, which was first investigated in the early 1950s and has been the subject of continuous investigation since then. Key to this idea is that radiation treatment of food is essentially isothermal, and it can therefore be applied to frozen foodstuffs as well as fresh, enabling bulk treatment. The dose of radiation varies according to the desired effect: 0.15 kGy is sufficient to prevent sprouting in potatoes, but up to 50 kGy may be required to destroy viruses in seafood and meat products. These higher doses of radiation are sufficient to bring about chemical changes in the composition of the food. Lipids are particularly susceptible to radiation damage, as the free radicals formed participate in a lipid peroxidation chain reaction.

Chapter Summary

- Deuterium exchange reactions play an important role in nuclear chemistry as it provides a means of investigating isotope exchange effects. The kinetic isotope effect is a useful technique in physical organic chemistry for the determination of reaction mechanisms.
- Stable isotope tracers are used in analytical chemistry for dilution analysis, while radioactive tracers can be used for detection of trace quantities of an analyte or for the determination of solubility constants.
- In neutron activation analysis, stable nuclei are rendered radioactive by neutron bombardment, which subsequently decay with unique spectroscopic signatures, which can be used to identify trace elements.
- Radiation chemistry is the study of the effects of radiation. Radiolysis is a particularly important phenomenon in biological systems, and also in the nuclear power industry. The measurement of absorbed radiation, known

as dosimetry, makes use of the chemical effects of ionizing radiation, particularly with regard to formation of oxidizing species.

Review Questions

(1) Identify the isotopologues and isotopomers from the following: CH_4, $^{13}CH_4$, $^{13}CO_2$, CH_3D, CH_2D_2, and CO_2.

(2) Rationalize the observation that D_3O^+ is a stronger Brønsted–Lowry acid than H_3O^+.

(3) Describe the difference between primary and secondary kinetic isotope effects and suggest some ways, in which this phenomenon is used in physical organic chemistry.

(4) How many carbon-11 atoms are there in $[U\text{-}^{13}C]$-methyl ethyl ketone?

(5) What would be the maximum specific activity $[U\text{-}^{11}C]$-methyl ethyl ketone, taking $t_{1/2}$ (carbon-11) = 20.33 min?

(6) Bromate ion containing 1.13% oxygen-18 was reacted with excess sulfurous acid:

$$BrO_3^- + 3HSO_3^- Br^- + 3SO_4^{2-} + 3H^+$$

The sulfate was found to contain 0.314% oxygen-18. What average number of oxygen atoms from the bromate was incorporated into the sulfate?

(7) A sample of river water was split into two aliquots: into one, $5\,\mu g$ of Al^{3+} was introduced and into the other a quantity of distilled water. On neutron activation analysis, the solution diluted with water gave a counting rate of 2315 cpm while that containing Al^{3+} gave a reading of 4197 cpm. Calculate the mass of aluminum in original sample.

(8) Tritium undergoes beta decay and releases a 5.65 keV beta particle. If a sample of tritiated water has a specific activity of 1 mC/g, what would be the estimated dose rate in Gy/min?

(9) Show that the correct units of absorbed dose are obtained by dimensional analysis of Eq. (9.24). NB: A is dimensionless.

(10) If a Fricke dosimeter is irradiated by 35 keV gamma rays from a ^{125}I source at an average absorbed dose of 17 Gy, what would be the corresponding change in absorbance? What effect would increasing the pathlength to 10 cm have?

Chapter 10

Nuclear Medicine

"Whatsoever house I may enter, my visit shall be for the convenience and advantage of the patient and I will willingly refrain from doing any injury."

The Hippocratic Oath

Nuclear medicine is a true interdisciplinary field drawing on the principles of nuclear physics, chemistry, and biology to diagnose and treat disease. On completion of this chapter and the associated questions, you should:

- Be able to apply the knowledge of radioactive decay and magnetic resonance to diagnostic imaging.
- Be able to describe the principles of the main divisions of radiation therapy.
- Understand basic concepts in radiopharmacy and the design of radiopharmaceuticals.

10.1 Introduction

Key Point: Nuclear medicine includes diagnostic use of radioisotopes and therapeutic use in radiation therapy. A contemporary issue is the production and supply of key radioisotopes.

The discovery of X-rays by Wilhelm Röntgen (1845–1923) in 1895 was integral in forming a link between physics and medicine. In fact, it was only one year later that Emil Grubbe (1875–1960) trialed the first use of X-rays in the treatment of breast cancer. In modern usage,

Table 10.1 Common Radionuclides Used in Medicine

Radionuclide	$t_{1/2}$	Decay	Gamma Energies (MeV)	Beta Energies (MeV)
Tracers				
^{3}H	12.3 y	β^-		0.018
^{14}C	5760 y	β^-		0.155
^{35}S	87.2 d	β^-		0.167
^{43}K	22 h	β^-	0.37; 0.61	0.83
^{51}Cr	27.8 d	EC	0.32	
^{57}Co	270 d	EC	0.122	
^{58}Co	71 d	β^+; EC	0.51; 0.81	0.485
^{59}Fe	45 d	β^-	1.10; 1.29	0.485
^{131}I	8.04 d	β^-	0.364	0.61
^{132}I	2.3 h	β^-	0.67; 0.78	2.12
^{99m}Tc	6 h	IT (γ)	0.14	
^{113m}In	90 m	IT(γ)	1.4	
^{75}Se	121 d	EC	0.14; 0.27	
^{68}Ga	68 m	β^+; EC	0.51	1.89
^{81m}Kr	13 s	IT(γ)	0.19	
^{133}Xe	5.3 d	β^-	0.081	0.34
Therapy				
^{32}P	14.3 d	β^-		1.71
^{90}Y	64.2 h	β^-		2.27
^{131}I	8.04 d	β^-	0.364	0.61
^{137}Cs	30 y	β^-	0.662	0.51
^{60}Co	5.3 y	β^-	1.17; 1.33	0.31

nuclear medicine is usually divided into two areas, diagnostic medical imaging and radiotherapy, both of which have strong links with other medical specialties, particularly oncology and neurology. About 95% of the radioisotopes produced for medical use are employed in medical imaging, with the remaining 5% being used in radiation therapy (Table 10.1). Such radioisotopes are generally divided into two categories:

(1) Proton-deficient radioisotopes, such as technetium-99m, which are usually produced in nuclear reactors.

(2) Neutron-deficient radioisotopes, such as fluorine-18, which are produced in proton accelerators (mostly cyclotrons).

The first production center for medicinal radioisotopes was the Oak Ridge National Laboratory in 1946, which supplied iodine-131 for trials in the treatment of thyroid cancer. Today, most of Europe's supply of medical radioisotopes comes from the Petten nuclear reactor facility in the Netherlands, while the Chalk River Laboratories supply in North America.

The main reactor at the Chalk River site has had several problems over the past 10 years or so. In 2007, the reactor was shut down for routine maintenance which took longer than anticipated, leading to a world shortage of radioisotopes, particularly the ubiquitous technetium-99m. In 2009, a heavy water leak led to a second shut down of the reactor, causing a further worldwide shortage of radioisotopes. To mitigate the effects of this in the future, steps are being taken to reduce the dependency on single suppliers of radioisotopes.

10.2 Diagnostic Imaging

Key Point: MRI is the leading imaging technique in modern medicine. However, more specialized technique using radioactive tracers provide functional information about various organ systems.

Diagnostic imaging from a clinical perspective is largely dominated by magnetic resonance imaging (MRI) which is based on ^{1}H-NMR. Other techniques, such as two-dimensional scintigraphy, single-photon emission computed tomography and positron emission tomography (PET) all employ radioisotopes. These radioisotopes are mostly all produced in cyclotrons, which is a growing trend in nuclear medicine, particularly with increasing use of diagnostic imaging. As is discussed later in the chapter, the radioisotope can be used in its free state, or bound to a larger molecule creating a radiopharmaceutical. These often have specific physical and chemical characteristics which lend themselves to a particular imaging technique. A range of common tracers is shown in Figure 10.1.

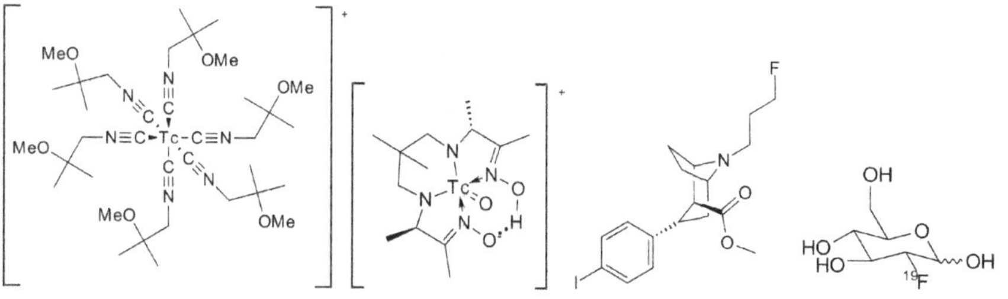

Figure 10.1 Common radioactive tracers in nuclear imaging. (a) Technetium-99m sestamibi (Cardiolite®); (b) technetium-99m exametazime (Ceretec®); (c) ioflupane; and (d) 2-deoxy-2-[^{18}F]fluoroglucose.

10.2.1 Magnetic resonance imaging

Key Point: MRI uses the principles of ^{1}H-NMR to acquire three-dimensional images of the internal structures of the body. Resolution is improved by manipulating the T_1 and T_2 relaxation times.

MRI, akin to its laboratory counterpart ^{1}H-NMR, is a non-invasive imaging technique which products a three-dimensional image of the body from the behavior of mobile protons in water, fats, and proteins. It has become a mainstay of diagnostic medicine, and continuous developments in MRI technology have reduced the running costs to such an extent that all hospitals in the developed world now have access to MRI facilities.

Although the underlying physics of the technique is identical to that of NMR, the hardware and signal acquisition process is markedly different. In a typical MRI scanner, the patient is placed on a sliding table which moves them into the MRI chamber (Figure 10.2). Here, they are placed in a magnetic field (1–3 T) and exposed to pulses (2–10 μs) of radiofrequency radiation (40–300 THz) from three directions. As the excited protons relax back to their ground state, they release radiofrequency radiation characteristic of their location. Measurement of the emitted radiation coupled with computed tomography software builds up a three-dimensional image. The contrast between different tissues in MRI arises from different signal intensities. For example, the signal due to protons present in bone

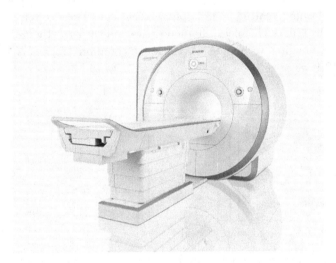

Figure 10.2 Siemens MAGNETOM spectra MRI scanner. This operates at 3 T and can be used for a variety of imaging applications.

arises at low frequencies, while that of muscle occurs at higher frequencies, which is consistent with the chemical environment of the protons.

The main source of contrast in MRI comes from differences in relaxation time, which leads to the T_1- and T_2-tissue-weighting factors. Protons in different tissues will have varying motilities, which will affect their relaxation pathway. The contrast in an image can be enhanced by altering the MRI pulse sequence to favor one of the relaxation pathways. Images which are T_1-weighted will show greater contrast between tissues with low T_1 and those that do not — for example, grey versus white matter in the brain. T_2 weighting is more sensitive to water content and can be used to locate regions of fluid accumulation, such as ascites.

MRI contrast can be further enhanced through administration of metal coordination complexes, so-called **contrast agents**, which are paramagnetic and greatly alter T_1 and T_2. The most common contrast agents are based on gadolinium complexes which have a large magnetic moment and are too large to enter healthy cells or cross the blood–brain barrier. The first gadolinium complex to be approved for clinical use was a complex of gadolinium(III) with diethylenetriaminepentacetate (DTPA), $[Gd(DTPA)H_2O]^{2-}$, which is used in the detection on intracranial lesions. Gadolinium compounds shorten the T_1 relaxation time by producing an oscillating magnetic field that increases the T_1 relaxation rate, increasing the contrast between tissues.

10.2.2 Two-dimensional scintigraphy

Key Point: Gamma-emitting radiopharmaceuticals are administered for functional studies of various organ systems. The image is captured using a type of scintillation counter known as a gamma camera.

Two-dimensional scintigraphy (2D-SCINT) is a versatile imaging technique employing gamma-emitting radioisotopes and a specialized **gamma camera** which collects two-dimensional images of the region under study. The general technique involves administration of a specific radiopharmaceutical, i.e., a radioisotope contained within a compound which specifically targets an organ or tissue. As the radioactive part of the drug disintegrates, the gamma emissions are monitored by a surrounding gamma camera, building up a composite image of the region under study.

The gamma camera employed in 2D-SCINT and other techniques is essentially a scintillation counter mounted on a movable gantry. The head of the camera consists of a large, flat crystal of sodium iodide doped with thallium, which is housed in a light-proof casing and backed with an array of photomultiplier tubes. As with other forms of scintillation counting, when a gamma ray photon is incident on the detector, an electron is ejected from the iodide forming an excited state; when it returns to a ground state, a photon of light is released. This is detected by the photomultiplier tube and the output is digitally reconstructed to give the two-dimensional image. To achieve spatial resolution, the gamma ray photons must be collimated, usually by a lead channel collimator, which gives an indication of the direction of the gamma ray photons and helps resolve the image.

Despite the relative sensitivity of a scintillation device to gamma rays, the resolution of 2D-SCINT is limited to about 1.8 cm. However, the advantage of 2D-SCINT and related techniques is that they can provide functional information to some extent. For example, in lung scintigraphy, a technetium-99m-DTPA tracer is used to diagnose chronic obstructive pulmonary disease. If there is reduced intake of air/tracer, the image will appear dark — these so-called "cold spots" are due to lower levels of gamma emission.

10.2.3 Single-photon emission computed tomography

Key Point: Technetium-99m radiopharmaceuticals are used for three-dimensional functional scans of cardiac or brain function.

Single-photon emission computed tomography (SPECT) is a functional imaging technique used to study blood flow through tissues and organs. This technique uses virtually the same gamma-emitting tracers and gamma camera as 2D-SCINT, but it then employs computed tomography to compile a three-dimensional image from the separate scans. A major use of SPECT is in functional studies of the heart, known as myocardial perfusion imaging, which are used to diagnose ischemic heart disease. In these investigations, the patient is administered a technetium-99m-labeled drug, such as ^{99m}Tc-tetrofosmin or ^{99m}Tc-sestamibi, which are lipophilic and can therefore pass into myocardial cells. In a typical investigation, the radio-labeled drug is administered, followed by stress testing in which the heart rate is increased. If the blood flow is reduced in any area of the heart (due to myocardial damage), dark patches appear on the SPECT image due to less gamma ray emission. Some cardiologists then request follow-up imaging some days after under resting conditions, especially if any abnormalities were detected in the initial investigation.

SPECT has also been used for functional brain imaging in patients with epilepsy, Alzheimer's disease or to map damage following a stroke. This application of SPECT uses ^{99m}Tc-exametazime as a tracer which is taken up by the brain in an amount proportional to blood flow. As blood flow is linked to brain activity, any areas of abnormally high or low activity will appear lighter or darker, respectively. This type of neuroimaging does not have particularly high resolution in comparison to MRI, but is more readily available and does not require access to a cyclotron for production of radioisotopes in-house.

10.2.4 Positron emission tomography

Key Point: Positron emitting radioisotopes are incorporated into various tracers. The most common tracer, 2-deoxy-2-[^{18}F]fluoroglucose, is used extensively to monitor the growth of tumors. The gamma rays released by pair annihilation are detected by a gamma camera.

PET is second only to MRI in terms of its use in diagnostic nuclear medicine. The technique relies on a range of positron-emitting radioisotopes with short half-lives, including fluorine-19 and rubidium-82. When the positron is released, it travels a short distance within the body before undergoing pair annihilation with an electron, producing two gamma ray photons. As with the other imaging techniques discussed, a form of scintillation counting is used to detect and quantify the photons. However, the localization of the gamma emission in PET is more complicated than in other techniques as the two gamma ray photons are emitted in opposite directions. To account for this, PET instrumentation only registers photon pairs that arrive within a few nanoseconds of each other, and then uses various algorithms to localize the point of origin.

The most common PET tracer is 2-deoxy-2-[^{18}F]fluoroglucose, (FDG), an analogue of glucose in which the 2'-hydroxyl group has been replaced by fluorine. Like glucose, FDG is taken up by cells through specific transporter proteins, after which it is phosphorylated by hexokinase. The FDG-phosphate cannot be metabolized further as it lacks a 2'-hydroxyl group and it accumulates intracellularly. However, when the fluorine-18 undergoes decay, oxygen-18 is formed and the subsequent gain of a proton produces a 2'-hydroxyl group, allowing the tracer to be metabolized largely as normal.

PET imaging is commonly used to monitor the growth of tumors, as these have increased levels of glucose uptake, and therefore accumulate

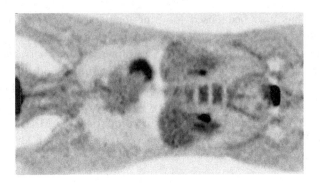

Figure 10.3 Normal distribution of FDG. PET image shows accumulation of FDG in the heart, liver, kidneys, bladder, and at the cerebral–cerebellar cortex at the base of the skull.

FDG. However, any tissues which have high levels of cellular respiration will appear dark on a PET image (Figure 10.3). It is also used to differentiate between Alzheimer's disease and other neurological conditions. This is based on the observation that the brain in patients with Alzheimer's disease has greatly decreased glucose uptake; the signal from FDG is therefore reduced in comparison to PET imaging of a healthy brain.

10.3 Radiation Therapy

Key Point: External or internal sources of radiation can be used to target diseased tissue with relative accuracy, minimizing collateral damage to adjacent, healthy tissues.

The use of ionizing radiation to destroy diseased cells probably began in the 1800s when X-rays were used to treat breast cancer and tuberculosis. The major drawback of these early procedures was the extensive collateral damage caused to healthy surrounding tissues by the radiation. While this remains a problem, modern techniques use precisely targeted radiation, avoiding secondary damage while destroying the cancer cells.

In ***external beam therapy***, beams of high-energy X-rays, protons or electrons are directed toward the affected region of the patient, usually after extensive imaging to plan the treatment. There are various subdivisions of external beam therapy which employ different radiation sources and delivery techniques. While most rely on linear accelerators to generate the radiation, the gamma knife uses a number of cobalt-60 sources to generate gamma rays which are directed toward a brain tumor. In ***brachytherapy***, small, radioactive implants are placed in the tissue adjacent to the tumor. From here, they deliver a constant dose of radiation within a relatively small region and have proved very effective in the treatment of breast, prostate and skin cancers. The most common radionuclides used are cobalt-60, caesium-137, and indium-192, although others are used in more specialist treatments. *Radioisotope therapy* involves directly administering a radioactive drug with chemical and physical characteristics which allow its accumulation in specific organs. The most common example of this is the use of iodine-131 ("radioiodine") in the treatment of thyroid disease; iodine is specifically absorbed by the thyroid tissues, eliminating the need for conjugation to a peptide ligand. A drug incorporating samarium-153 is similarly used for the treatment of bone metastases. When administered intravenously, the drug is preferentially absorbed by rapidly-dividing cancer cells in the bone, sparing healthy bone.

10.4 Radiopharmaceuticals

Key Point: Production of tracers and drugs for use in nuclear medicine involves preparation of the radionuclide (*via* nuclear reactor or cyclotron) and incorporation into a large molecule for delivery.

The preparation of radiopharmaceuticals involves two main steps: the preparation/isolation of the radionuclide and the conversion of the radionuclide into a dosage form, usually through complex formation. Although the potential range of radiopharmaceuticals is vast, the regulatory and economic considerations have limited the use of these drugs to a relatively small selection. The seemingly ubiquitous technetium-99m-based drugs and 2-deoxy-2-[^{18}F]fluoroglucose typify much of the chemistry involved.

10.4.1 Production of radioisotopes in nuclear reactors

Key Point: Radioisotopes can be produced by bombardment of stable nuclei by thermal neutrons in nuclear reactors, or by charged particles in a cyclotron.

Nuclear reactors can be used to produce proton-deficient radioisotopes through bombardment of stable nuclei with thermal neutrons. Alternatively, nuclear fission of a heavy element can provide a variety of useful radionuclides. In the first case, thermal neutrons can mediate two types of reaction:

$$\substack{Y \\ Z}A + \substack{0 \\ 1}n \rightarrow \substack{Y+1 \\ Z}A + \gamma \quad \text{e.g.,} \quad {}^{98}\text{Mo}(n, \gamma){}^{99}\text{Mo} \qquad (10.1)$$

$$\substack{Y \\ Z}A + \substack{0 \\ 1}n \rightarrow \substack{Y \\ Z-1}A + \substack{1 \\ 0}p \quad \text{e.g.,} \quad {}^{14}\text{N}(n, \gamma){}^{14}\text{C} \qquad (10.2)$$

The reaction represented by Eq. (10.1) is common and will tend to yield products of low specific activity, while that shown in Eq. (10.2) gives products with high specific activity and are carrier free. Alternatively, fission reactions of the type

$$\substack{235 \\ 92}\text{U} + \substack{0 \\ 1}n \rightarrow \text{fission products} + \nu \substack{0 \\ 1}n \qquad (10.3)$$

can be used to produce radioisotopes from any heavy element, although uranium-235 is the most commonly used. Of course, the major problem with reactions of this type is separation of the desired radioisotope from the other fission products.

By far the most used radionuclide in modern nuclear medicine is technetium-99m, a decay product of molybdenum-99. The most common source of molybdium-99 is from fission of uranium-235, which produces a range of fission products, including about 6% molybdium-99 (Figure 10.4). After the fission is complete, and depending on the composition of the uranium target, the molybdenum content may be recovered by acid or base treatment. Most targets are composed of uranium and aluminum and these are processed by base treatment. Once the target has been cooled in water, it is transported to a processing facility where it is dissolved in

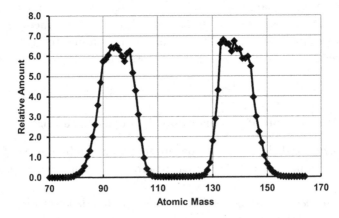

Figure 10.4 Fission products of uranium-235. The range of fission products produced in a nuclear reactor is vast; however, these can be separated from molybdenum-100 by physical and chemical methods.

sodium hydroxide to form (*inter alia*) sodium molybdate (Na_2MoO_4). This solution is filtered, purified by ion exchange and the molybdate content recovered by passing down a column of alumina. Due to the short half-life of molybdenum-99, only about 10% of the purified produced reaches the end user. For this reason, the radioisotope (or its technetium-99*m* decay product) cannot be stored for any appreciable length of time.

When the molybdenum-99 has been purified, it is commonly used to manufacture a "moly cow" for delivery to radiopharmacy departments. The basic layout of a wet-column moly cow is shown in Figure 10.5. First, an alumina column is prepared which is loaded with a solution of sodium molybdate. When the pH is lowered, the molybdate anion polymerizes in the presence of alumina to form a hetropolymer, $Al[Mo_6O_{24}]^{9-}$. Although the physical quantity of molybdenum on the column will be low, the specific activity is high (*ca.* 370 TBq per gram). The molybdenum will decay to technetium-99*m* (87%) and technetium-99 (13%), reaching a transient equilibrium. The pertechnetate ($^{99m}TcO_4^-$) is eluted from the column by a suitable solution, usually saline in a clinical setting, and passed through a micro filter to remove harmful organisms.

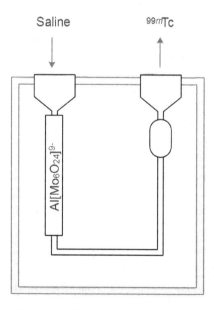

Figure 10.5 Representation of a moly cow. A saline solution is introduced onto the column, which is pulled through by an evacuated vial attached to the other port. The eluted $^{99m}TcO_4^-$ solution is passed through a sterile filter before use.

The eluate from a $Al[Mo_6O_{24}]^{9-}$ column will also contain $^{99}TcO_4^-$ as a carrier, which will to some extent dilute the radioisotope depending on how many half-lives have passed. The mole fraction, F, of the eluate can be calculated at any time, t, from

$$F = \frac{0.87(e^{-\lambda_1 t} - e^{-\lambda_2 t})}{(\lambda_2 \lambda_1)(1 - e^{-\lambda_1 t})} \tag{10.4}$$

where λ_1 and λ_2 are the decay constants for molybdenum-99 and technetium-99m, respectively. The factor 0.87 takes into account the fraction of molybdenum-99 that decays to technetium-99m.

10.4.2 Production of radioisotopes in cyclotrons

Key Point: Cyclotrons mostly produce positron-emitted radionuclides for use in imaging studies. Modern techniques may also prove useful for production of technetium-99m using medical cyclotrons.

As reviewed in Chapter 8, cyclotrons produce a stream of charged particles, constrained to follow a circular path by magnetic fields and accelerated by a potential difference. Eventually, the charged particles emerge and strike a target, bringing about a nuclear transformation. The use of on-site cyclotrons to produce radioisotopes was pioneered by the Hammersmith Hospital in London in 1955 and the technology was later commercialized by Phillips in 1966.

For the most part, cyclotrons are employed to produce positron-emitting radioisotopes with relatively short half-lives (e.g., carbon-11, nitrogen-14, oxygen-15, and fluorine-18). Probably the most prevalent positron emitter is fluorine-18, which may be incorporated into a variety of radiopharmaceuticals. Although fluorine-18 can be produced by a variety of nuclear reactions (Table 10.2), for all practical purposes, it is achieved by proton bombardment of oxygen-18-enriched water. The proton beam is usually directed at a titanium target vial containing oxygen-18-enriched water at high pressure. At the end of bombardment, the aqueous $[^{18}F]$

Table 10.2 Production of Fluorine-18

Reaction	Target	Product	SA (GBq/μmol)
$^{18}O(p,n)^{18}F$	$H_2^{18}O$	$[^{18}F]$fluoride(aq)	40×10^3
$^{16}O(^3He,p)^{18}F$	H_2O	$[^{18}F]$fluoride(aq)	40×10^3
$^{20}Ne(d,\alpha)^{18}F$	Ne/F_2	$[^{18}F]F_2$	*ca.* 0.04–0.40
$^{18}O(p,n)^{18}F$	$^{18}O_2/Kr/F_2$	$[^{18}F]F_2$	*ca.* 0.04–0.40

fluoride is pushed out of the vial by a stream of argon. The fluorine content can then be recovered by ion exchange chromatography and then incorporated into the pharmaceutical by standard chemical techniques.

There is potentially a means of producing proton-deficient radioisotopes using cyclotron technology. In the 1970s, it was shown that molybdium-99 could be produced from molybdenum-100 using linear accelerators. When high-energy electrons are directed a target, they produce bremsstrahlung photons which can then be focused on an enriched molybdenum-100 target, ejecting a neutron and producing molybdenum-99:

$$^{100}_{42}\text{Mo} + \gamma \rightarrow \,^{99}_{42}\text{Mo} + \,^{1}_{0}n \qquad (10.5)$$

The molybdenum-99 can then be used to produce moly cows as previously described. In 2015, Canadian researchers reported an efficient, relatively cost-effective means of producing technetium-99m directly from molybdenum-100 using cyclotrons:

$$^{100}_{42}\text{Mo} + p \rightarrow \,^{99m}_{43}\text{Tc} + 2\,^{1}_{0}n \qquad (10.6)$$

In this approach, a target was first prepared by electrodepositing molybdenum-100 on thin aluminum sheet. This was then irradiated by a beam of protons to produce the technetium-99m which can then be separated from traces of molybdenum and aluminum by ion exchange chromatography. The molybdate eluate from the purification process can be recycled back to molybdenum-100, minimizing waste and reducing the cost of the process.

10.4.3 Production of radiopharmaceuticals

Key Point: Preparation of technetium-99m compounds involves the reduction of pertechnetate to technetium(V) by tin. [^{18}F] fluoridation proceeds by well-established reactions such as the Walden inversion.

Although some radionuclides can be used directly, must be incorporated in larger molecules to increase their uptake and reduce harmful side effects. Much of this relies on traditional techniques in coordination chemistry and organic chemistry.

Technetium-99m remains the most widely used radionuclide for diagnostic imaging. This is partly due to its favorable chemical and radiochemical characteristics, but also the ease with which it can be obtained. These compounds can be classified as being Tc-essential (they depend on technetium's coordination chemistry) or Tc-tagged (technetium-99m acts as a radiolabel). The formation of either type of compound will (usually) involve pertechnetate, as this is what is eluted from a moly cow. The pertechnetate ion is one of the most thermodynamically stable forms of the element and is in some ways analogous to permanganate, except that the TcO_4^-/TcO_2 redox couple has a much lower reduction potential:

$$TcO_4^- + 4H^+ + 3e^- \rightleftharpoons TcO_2 + 2H_2O \quad E^\circ = +0.747\,V \qquad (10.7)$$

In order for a stable coordination complex to form, pertechnetate must first be reduced to a lower oxidation state. Although several reducing agents are suitable, stannous ion is the by far the most widely used. In acidic solution, the stannous ion is readily oxidized to tin(VI):

$$Sn^{2+} \rightarrow Sn^{4+} + 2e^- \quad E^\circ = -0.15\,V \qquad (10.8)$$

The redox chemistry of tin is complicated and its oxidation may occur in a sequence of reactions. With regard to reduction of pertechnetate to technetium(V) by stannous chloride, the dominant process in acid is believed to be

$$TcO_4^- + SnCl_3^- \rightleftharpoons TcO_3^- + OSnCl_3^- \qquad (10.9)$$

However, the technetium may be further reduced to an insoluble oxide, $TcO_2.xH_2O$, which is undesirable. To prevent this, the reduction is performed in the presence of a ligand which stabilizes the lower oxidation state. A common chelation pattern in technetium-99m radiopharmaceuticals is shown in Figure 10.6. The three chelate rings are typically five- or six-membered, coordinating through amines, phosphines, thiolates, etc. The

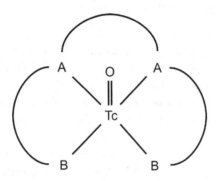

Figure 10.6 Arrangement of ligands in a Technetium-99m radiopharmaceutical. The atoms at positions A and B usually belong to neutral coordinating groups (e.g., amines).

ring backbones can be substituted by a variety of functional groups to vary the physicochemical properties of the drug.

The most common coordination compounds used for brain imaging are Ceretec® and Neurolite®, which are neutral, moderately lipophilic square-pyramidal complexes of technetium-99m. As these compounds are uncharged and have a relatively low molecular mass, they can pass through the blood–brain barrier and accumulate in the brain. Similar complexes are used for cardiac imaging: Myoview® and Cardiolite® are octahedral complexes of technetium-99m which carry a single positive charge. This latter feature enables the complexes to be taken up by cardiac muscle, possibly through an attraction to mitochondria within the muscle cells.

Tc-tagged radiopharmaceuticals incorporate a technetium-99m label and a biomolecule specific to the target organ. Conjugation of the inorganic label with the bioorganic group usually involves a linker. An example of this chemistry is found in the combination of technetium-99m with progesterone, linked by a phenyl group. This drug specifically targets the progesterone receptor which is expressed by a number of tissues include those of the prostate gland.

The production of fluorine-18-containing radiopharmaceuticals is based on electrophilic or nucleophilic reactions with fluorine. In the first case, direct combination of an electron-rich substrate with $[^{18}F]F_2$ can produce useful products, provided that regioselectivity is not a concern. The first commercial production of FDG was achieved by direct *cis*-addition of $[^{18}F]F_2$ to tri-*O*-acetyl-*D*-glucal, producing two fluoridated epimers which could be subsequently resolved. The specific activity and radiochemical yield of this procedure was fairly unremarkable.

Nucleophilic ^{18}F-fluoridation using $[^{18}F]$fluoride is the current method for preparation of FDG. When it is prepared, $[^{18}F]$fluoride is highly solvated in aqueous solution, which reduces its nucleophilicity. To improve this, the fluoride is first passed down an anion exchange column and then eluted with aqueous potassium carbonate. This is combined with a phase transfer catalyst, Kryptofix 2.2.2, and successfully dried by azeotropic distillation. This is then refluxed with 1,3,4,6-tetra-*O*-acetyl-2-*O*-trifluoromethane-sulfonyl-*β*-*D*-mannopyranose to produce FDG. The reaction proceeds as a Walden inversion and produces an enantiomerically pure derivative. The work-up consists of passing the derivative through several C-18 Sep-Pak columns, followed by acid hydrolysis and further purification on a C-18 column. These extra purification steps are necessary as Kryptofix 2.2.2 is cytotoxic.

Chapter Summary

- Nuclear medicine uses radionuclides for diagnostic imaging and for radiation therapy. Both applications are used extensively in the management of cancer.
- The main forms of diagnostic imaging which use radioactive tracers are: two-dimensional scintigraphy, single-photon emission computed tomography and PET. These provide functional information about an organ system, which can be used to diagnose specific illnesses.
- In radiation therapy, the ionizing radiation produced by the decay of select radionuclides destroys the DNA of diseased cells, bringing out programmed cell death.
- Production of radiopharmaceuticals requires a source of radionuclide, produced in a nuclear reactor or by a medical cyclotron. These are then incorporated into a chelate, which targets specific areas of the body. In some instances, just the radionuclide itself is required — for example, xenon-133.
- The production of key radionuclides like technetium-99m is an important issue in nuclear medicine. Emerging technologies include modifying medical cyclotrons to produce technetium-99m directly from molybdenum-100.

Review Questions

(1) Suggest why nuclear reactors are primarily used for production of proton-deficient radioisotopes.

(2) Explain why alpha emitting radioisotopes would not be suitable for use in radiotracers.

(3) Summarize the use of gamma-emitting radioisotopes in nuclear imaging and radiation therapy.

(4) Suggest why the specific activity of radionuclides produced in a nuclear reactor is higher than that produced by the (n, γ) reaction.

(5) Would the specific activity be high or low for radioisotopes with a long half-life?

(6) Which relaxation mechanism would gadolinium complexes affect? Explain your answer.

(7) Distinguish between brachytherapy and radioisotope therapy.

(8) Preparations containing ^{99m}Tc should not be exposed to oxygen or oxidizing agents. Suggest an explanation for this.

(9) A moly cow was prepared so that it would have 96.2 GBq of activity on 12 noon on Wednesday. What would be the activity of the eluate at 8 AM on Friday?

(10) Comment on the suitability of direct fluoridation with ^{18}F$_2$ in the production of radiopharmaceuticals. Do you think the standard electrode potential of fluorine ($E = +2.87$) is significant?

Chapter 11

Chemistry of the *f*-Block Elements

"Chemistry begins in the stars. The stars are the source of the chemical elements, which are the building blocks of matter and the core of our subject."

P. Atkins

The radioactive isotopes of elements have more or less the same chemical reactivity as their stable counterparts. The inorganic chemistry of *f*-block elements has relevance to the nuclear power industry and nuclear medicine. On completion of this chapter and the associated questions, you should:

- Be able to follow trends across the *f*-block and relate these to electronic structure and oxidation state.
- Understand the general properties of the lanthanoids and the unusual nature of promethium.
- Be able to apply knowledge of nuclear chemistry to the synthesis of the man-made elements and understand that their separation is based on knowledge of their chemical properties.

11.1 Introduction

Key Point: The lanthanoids (Period 6) and actinoids (Period 7) have *f*-orbital valence electrons. All actinoids after uranium are regarded as man-made elements.

The elements occupying what has come to be known as the *f*-block of the Periodic table have endured a number of name changes over the years.

Traditionally, those elements in Period 6 (curium to lutetium) were known as the "rare Earth elements," although they are actually fairly abundant in the Earth's crust (except for promethium which has no stable isotopes). In most instances, these are now known as the *lanthanoids* (Ln) while those elements in Period 7 (thorium to lawrencium) are collectively the *actinoids* (Ac).

The majority of the lanthanoids are obtained from mineral sources, such as monazite and xenotime, where they are usually found in the Ln(III) oxidation state. This makes extraction difficult, and usually extensive liquid–liquid extraction or ion-exchange procedures are required. Only two of the actinoids, thorium and uranium, have sufficiently long half-lives to ensure that significant amounts have persisted since their formation (they would have been formed in the supernova explosion which created our solar system ∼6.5 billion years ago.). Thorium can be extracted from the mineral thorite, $ThSiO_4$, while uranium is obtained from uranite, UO_2, or pitchblende, U_3O_8.

Purified lanthanoids are soft, white metals which react with steam and dilute acids. They are poorer conductors of heat and electricity than the *d*-block metals. Their value is in the compounds they form; their electronic structure allows optical *f–f* transitions, which are used in cathode-ray tubes (europium) and solid-state lasers (neodymium, samarium, and holmium). The actinoids are dense elements, but aside from the properties of uranium and plutonium, very little is known of their chemistry. While the focus of this chapter will be on actinoid chemistry, for the sake of having a more complete discussion, a brief treatment of the lanthanoids will also be given.

11.2 Lanthanoid Chemistry

Key Point: The lanthanoids are the first row of the *f*-block. Much of their chemistry is based on the Ln(III) oxidation state.

Moving across Period 6, electrons are added to the 4f orbitals (Figure 11.1), giving a general electronic configuration of $[Xe]6s^2 4f^x$, with the exception of gadolinium which is more stable as $[Xe]6s^2 5d^1 4f^7$. The *f*-orbitals are relatively diffuse and tightly bound to the nucleus; therefore they do not take part readily in chemical reactions. Therefore, during ion formation, the 4f orbitals are more stable than the 5d or 6s, favoring M^{3+} ions. These ions are typically colored due to electronic transitions between *f*-orbitals. The ionic radius decreases across from La to Lu, due in part to the increasing effective nuclear charge, but also relativistic effects. This

(a)   (b)   (c)

Figure 11.1 Cubic symmetry f-orbitals. As with all atomic orbitals, the f-orbitals are linear combinations of those predicted for hydrogenic atoms. (a) f_x^3, f_y^3, or f_z^3; (b) $f_x(y^2 - z^2)$, $f_y(z^2 - x^2)$, or $f_z(x^2 - y^2)$; (c) f_{xyz}.

is usually referred to as the lanthanide contraction and is characteristic of these elements.

The reduction potentials of the lanthanoids are very similar, with europium being typically anomalous; consequently the reactivity series for the lanthanoids is:

$$\underline{\text{Eu}}, \text{La}, \text{Ce}, \text{Pr}, \text{Nd}, \text{Sm}, \underline{\text{Yb}}, \text{Gd}, \text{Tb}, \underline{\text{Y}}, \text{Ho}, \text{Er}, \text{Tm}, \text{Lu}$$

Those underlined deviate from their periodic order. Cerium is atypical in that the Ce(IV) oxidation state has extensive chemistry. The Ce^{4+} ion is widely used as a one-electron oxidizing agent in analytical and organic chemistry. All Ln^{3+} ions are hard Lewis acids and readily form complexes through fluorine- and oxygen-containing ligands.

The Ln(III) ions occupy volumes 4–5 times greater than 3d-metal ions and therefore even the binary oxides have reasonably complex structures. Chemically, they resemble calcium oxide; they readily absorb carbon dioxide and water from the atmosphere, forming carbonates and hydroxides, respectively. Addition of fluoride ion to aqueous solutions of the lanthanoids produces insoluble fluorides, which is a characteristic test for these elements. The chlorides can be formed through reaction of the oxide with ammonium chloride at 300°C:

$$M_2O_3 + 6NH_4Cl \rightarrow 2MCl_3 + 3H_2O + 6NH_3 \qquad (11.1)$$

The aqua ions have coordination numbers greater than six, such as $[Nd(H_2O)_9]^{3+}$ and they hydrolyze in water with increasing ease as ionic radius decreases:

$$[M(H_2O)_n]^{3+} + H_2O \rightleftharpoons [M(OH)(H_2O)_{n-1}]^{2+} + H_2O^+ \qquad (11.2)$$

Lanthanoids form a variety of organometallic compounds, some of which have commercial applications. For example, neodymium complexes are used in the Ziegler–Natta polymerization of alkenes.

Promethium is the only lanthanoid which has no stable isotopes; it also is the only lanthanoid which does not follow the lanthanoid contraction. It occurs naturally through alpha decay of europium-151 or by spontaneous fission of uranium-238. Consequently, most natural sources are found in uranium ores. Of the 36 reported isotopes of promethium, the most stable is promethium-145 ($t_{1/2}$=17.7 year), which is a soft beta emitter. Promethium salts luminesce in the dark and the metal dissolves in aqueous sulfuric acid to form a pink complex:

$$2Pm + 3H_2SO_4 \rightleftharpoons 2[Pm(H_2O)_9]^{3+} + 3SO_4^{2-} + 3H_2 \qquad (11.3)$$

Promethium(III) chloride is still used as a component in some luminous paints, but the majority of its applications arise from its radioactive nature, mainly in thickness gauges.

11.3 The Actinoids

Key Point: All actinoids are radioactive; those after uranium in the series are produced by neutron or heavy atom bombardment.

Aside from uranium and thorium, which are obtained from mineral sources, the remaining actinoids are man-made, a process which began during the early stages of the Manhattan Project. The first new elements, neptunium and plutonium, were discovered through bombardment of uranium-238:

$$^{238}_{92}U \xrightarrow{n,\gamma} {}^{239}_{92}U \xrightarrow[23.5\,\text{min}]{\beta^-} {}^{239}_{93}Np \xrightarrow[2.35\,\text{days}]{\beta^-} {}^{239}_{94}Pu \qquad (11.4)$$

As the half-life of plutonium-239 is 24360 years, this process cannot be used to isolate neptunium-239. Instead, a more stable isotope (neptunium-237, $t_{1/2} = 2.2 \times 10^6$ years) can be prepared by bombardment of uranium-235:

$$^{235}_{92}U \xrightarrow{2n,\gamma} {}^{237}_{92}U \xrightarrow[6.75\,\text{min}]{\beta^-} {}^{237}_{93}Np \qquad (11.5)$$

Americium, curium, berkelium, californium, and einsteinium are prepared by successive neutron capture using plutonium-239 as the starting material;

for example, americium-241:

$$^{239}_{94}\text{Pu} \xrightarrow{n,\gamma} {}^{240}_{94}\text{Pu} \xrightarrow{n,\gamma} {}^{239}_{94}\text{Pu} \xrightarrow[13.2\,\text{years}]{\beta^-} {}^{241}_{95}\text{Am} \qquad (11.6)$$

The remaining actinides are prepared by bombardment of plutonium, americium, or curium with accelerated boron, carbon, or nitrogen ions.

11.3.1 Oxidation states and general redox behavior

Key Point: The Ac(III) oxidation state dominates actinoid chemistry. The aqua ions are prone to disproportionation in a pH-dependent manner.

Unlike the lanthanoids, the elements in Period 7 do not exhibit much chemical uniformity, particularly with respect to their oxidation states; in fact, the variety of oxidation states in the actinoid series is reminiscent of the first transition series (Table 11.1). This variation in oxidation state is probably because the 5f, 6d, and 7s orbitals have similar energy. The electrons in the 6d and 7s orbitals are more loosely bound that those in the 5f orbitals. Thus, it is easy to see that actinium will lose its $6d7s^2$ electrons to give the Ac^{3+} ion, and similarly thorium forms that Th^{4+} ion through loss of its $6d^27s^2$ electrons. For subsequent elements, the 5f orbitals have a greater contribution and this makes general trends impossible to define.

The An(III) oxidation state is common to all actinoids except thorium and protactinium, and has chemistry similar to that of the lanthanoids. Thorium exists principally in the Th(IV) oxidation state and protactinium prefers the Pa(V) state.

Dioxo ions of the type MO_2^+ and MO_2^{2+} are found in solution and their stability depends on their tendency to undergo disproportionation. The

Table 11.1 Oxidation States of the Actinoids. The Favored (Most Stable) Oxidation State of Each Element is Shown in Bold

Ac	Th	Pa	U	Np	Pu	Am	Cm	Bk	Cf	Es	Fm	Md	No	Lr
						2			2	2	2	2	**2**	
3	3	3	3	3	3	**3**	**3**	**3**	**3**	**3**	**3**	**3**	3	3
	4	4	4	4	4	4	4	4						
		5	5	5	5	5			5?					
			6	6	6	6								
				7	7									

PuO_2^+ ion disproportionates to give a variety of species:

$$2PuO_2^+ + 4H^+ \rightleftharpoons Pu^{4+} + PuO_2^{2+} + 2H_2O \qquad (11.7)$$

$$2PuO_2^+ + Pu^{4+} \rightleftharpoons Pu^{4+} + PuO_2^{2+} + Pu^{3+} \qquad (11.8)$$

Despite the disproportionation of PuO_2^+, it is observed in oceanic waters, mostly probably due to the formation of a bicarbonate complex which stabilizes it and prevents rapid disproportionation. In fact, the redox behavior of the actinoids is significantly pH dependent, which is important when considering the state of the actinoids in the environment. For example, the plutonium(III) and neptunium(III) oxidation states are favored at pH 8, which is the approximate pH of natural waters and compounds of these ions can be detected in the sediment. In general, it is found that hydrolysis

$$An^{+n} + mH_2O \rightleftharpoons OAn(OH)_m^{+(n+m)} + mH^+ \qquad (11.9)$$

follows the order $An^{4+} > An_2^{2+} > An^{3+} > AnO_2^+$ with increasing pH. Thus, ions such as PuO_2^+ and UO_2^+ would be more stable at lower pH.

11.3.2 Actinium, thorium and protactinium

Key Point: The first three elements occur in varying amounts in nature. Actinium-225 is used in emerging radiotherapy treatments.

Actinium occurs in trace amounts in uranium and thorium ores, and can be made in milligram quantities from neutron bombardment of radium-226. The metal is slivery in appearance and it forms oxides and halides in much the same way as lanthanoids. The scarcity of actinium means that it has found little wide-spread application in industry. A novel use of the actinium-225 isotope is in ***targeted alpha therapy***, where actinium can be administered in an antibody conjugated form (Figure 11.2). The localized release of alpha particles damages the DNA in the target cell, promoting cell death.

Thorium is widely distributed across a number of ores and is extracted from monazite sand by digestion with sodium hydroxide. When the pH is lowered, thorium hydroxide is precipitated out and recovered by solvent extraction using tributyl phosphate in kerosene. Thorium is mostly used as its nitrate, $Th(NO_3)_4.5H_2O$, which is soluble in water, alcohols, ketones, and esters. On heating, the nitrate gives thorium dioxide, which emits intense blue light when hot; this was used in combination with cerium

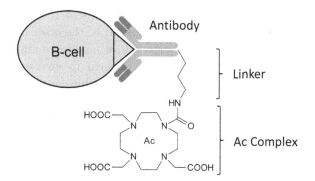

Figure 11.2 Targeted alpha therapy. Actinium is chelated by an appropriate macrocyclic ligand which is conjugated to an antibody. The antibody recognizes specific epitopes on the surface of a B-lymphocyte, targeting the alpha particles. These destroy the DNA of the damaged cell.

dioxide in gas mantles. Thorium oxide has the highest melting point of any oxide and it is used in high-temperature laboratory crucibles.

Protactinium is one of the most expensive naturally occurring metals and is found in uranium ore. Its major oxidation states give rise to a range of oxides, hydroxides, and halides. Aside from its radioactivity, protactinium is toxic and this has limited its use beyond research. Measurement of the protactinium-231 to thorium-230 ratio can be used in radiometric dating of ocean sediment, which is used to model the formation of minerals.

11.3.3 Uranium

Key Point: Uranium is the central component of the nuclear power industry. Formation of uranyl nitrate, $UO_2(NO_3)_2.nH_2O$, is the first step in the purification of uranium yellowcake.

The discovery of naturally occurring uranium is credited to Martin Klaproth (1743–1817) who obtained sodium diuranate by dissolving pitchblende in nitric acid and neutralizing with sodium hydroxide. The yellow salt was used extensively in the production of uranium glass and was a component of early yellowcake uranium mixes. In nature, uranium exists primarily as three isotopes: uranium-238 (99.27%), uranium-235 (0.72%), and uranium-234 (0.005%). All three are alpha emitters, but it is the uranium-235 isotope which is fissionable. One of the achievements of the Manhattan Project was developing a gas diffusion method for enrichment of uranium-235 (see Chapter 12).

Uranium is an abundant element and is present in its common ore, pitchblende, as uranium dioxide. Aside from this, the other chief oxides of uranium are the trioxide, UO_3 (orange–yellow) and triuranium octaoxide, U_3O_8 (black). All oxides react with nitric acid to give uranyl nitrate, $UO_2(NO_3)_2.nH_2O$, which is the first step in the purification process. The uranyl nitrate is then reacted with ammonia to give ammonium diuranate which is reduced by hydrogen gas to give pure uranium dioxide. This is then reacted with hydrofluoric acid and fluorine to give uranium hexafluoride:

$$UO_2 + HF \rightarrow UF_4 + 2H_2O \qquad (11.10)$$

$$UF_4 + F_2 \rightarrow UF_6 \qquad (11.11)$$

Uranium hexafluoride has a high vapor pressure, which is used in the commercial enrichment process.

The uranyl ion, or dioxouranium(VI), is linear and can form a variety of complexes with different coordination numbers (Figure 11.3). In aqueous solution, it behaves like a weak acid with a pK_a of *ca.* 4.2:

$$[UO_2(H_2O)_4]^{2+} \rightleftharpoons [UO_2(H_2O)_3(OH)]^+ + H^+ \qquad (11.12)$$

The uranyl ion, which is yellow in aqueous solution, can be easily reduced to uranium(IV) (green) by mild reducing agents such as zinc; reduction to uranium(III) (red–brown) can be achieved using a Jones reductor

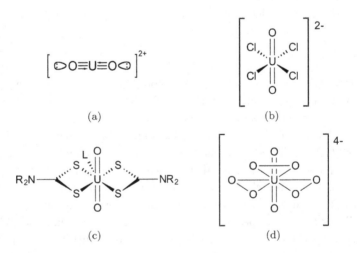

(a)   (b)

(c)   (d)

Figure 11.3 Uranyl complexes. The uranyl ion (a) can form a variety of complexes (b)–(d) exhibiting a coordination of nine. Other actinoid complexes can have coordination numbers as high as 12.

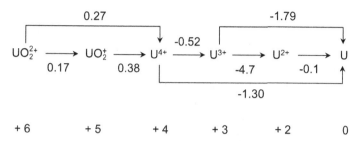

Figure 11.4 Latimer diagram for uranium ions. The dioxouranium ion can be successively reduced to metallic uranium by selection of appropriate reducing agents. All values are quoted in volts. Data from L. R. Moss, *The Chemistry of the Actinide Elements*, Vol. 2, Chapman & Hill, London (1986).

(Figure 11.4). The colors of the uranium ions in aqueous solution are due to electronic transitions between f-orbitals. Technically, $f-f$ transitions are forbidden by the Laporte rule, but the selection rule is relaxed due to the Jahn–Teller effect (the symmetry of the ion is distorted by the surrounding ligands). A similar explanation applies for other actinoid aqua ions.

11.3.4 Neptunium, plutonium, and americium

Key Point: The first three elements after uranium were synthesized as part of the Manhattan Project research. Some trace amounts of plutonium are found in natural fusion reactors.

Neptunium was the first of the so-called transuranium elements to be discovered and was synthesized in 1940 at the Berkeley Radiation Laboratory, Eq. (11.5). Although the neptunium present at the formation of the solar system will have long since decayed, trace amounts of neptunium-237 are found as decay products of uranium. Most neptunium (and plutonium) found in the environment today is due to early nuclear weapons testing.

As was the case with neptunium, trace amounts of plutonium can be found in nature, particularly plutonium-239 which may be formed in natural fusion reactors such as the Oklo site in Africa. Plutonium-238 is an alpha emitter and is used in radioisotope thermoelectric generators. Plutonium-239 and plutonium-241 are fissile isotopes, while plutonium-240 undergoes spontaneous fission. The presence of plutonium-240 in a quantity of plutonium determines it "grade," that is, weapons-grade, fuel-grade, or reactor-grade.

Americium can be found in uranium deposits, but the vast majority of this element is formed in nuclear reactors. Americium dioxide, AmO_2, is the main form of the element used in commerce, mostly in smoke detectors. It is prepared by dissolving americium in hydrochloric acid, neutralizing the acid using ammonia, and then precipitating crystals of americium oxalate by addition of oxalic acid. These are filtered, washed, and calcined to form americium dioxide.

Much of the chemistry of neptunium, plutonium, and americium is broadly similar to that of uranium and can be exploited to separate mixtures of the elements. For example, the relative stability of the major oxo ions is

$$UO_2^{2+} > NpO_2^{2+} > PuO_2^{2+} > AmO_2^{2+}$$

By carefully selecting the conditions, a solution containing the elements in different oxidation states can be obtained. These can then be separated by precipitation or solvent extraction. Chemistry of this type is exploited in the tributyl phosphate solvent extraction cycle, such as that employed to separate uranium and plutonium (Figure 11.5). The solvent forms a complex with the uranium and plutonium nitrates, $AcO_2(NO_3)_2.2TBP$, while americium and curium remain in the aqueous phase. Plutonium is then separated from uranium by careful reduction of the Pu^{4+} ion to Pu^{3+} and back-extraction of the uranium into water. The plutonium is then

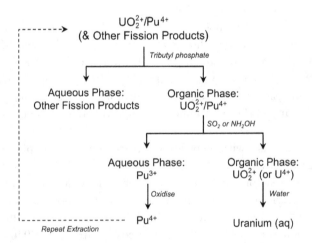

Figure 11.5 Tributyl phospate extraction. A solution of the fission products is prepared in nitric acid and extracted with tributyl phosphate in kerosene.

reoxidized and the extraction procedure repeated to achieve the maximum yield.

11.3.5 Late actinoids

Key Point: Curium — Lawrencium are entirely man-made elements, first detected following thermonuclear testing. They have similar chemical reactivity and exist chiefly in the Ac(III) oxidation state.

Following the initial work at Berkeley, another eight elements after americium were discovered. Curium was produced by Seaborg *et al.* in the mid-1940s by alpha particle bombardment of plutonium:

$$^{239}_{94}\text{Pu} + ^{4}_{2}\text{He} \longrightarrow ^{242}_{96}\text{Cm} + ^{1}_{0}\text{n} \tag{11.13}$$

Curium-242 decays through alpha emission ($t_{1/2}$=163 days) and spontaneous fission. Later, Seaborg's group irradiated americium-241 and curium-242 with alpha particles, producing a further two new elements, berkelium and californium:

$$^{241}_{95}\text{Am} + ^{4}_{2}\text{He} \longrightarrow ^{243}_{97}\text{Bk} + 2^{1}_{0}\text{n} \tag{11.14}$$

$$^{242}_{96}\text{Cm} + ^{4}_{2}\text{He} \longrightarrow ^{245}_{98}\text{Cf} + ^{1}_{0}\text{n} \tag{11.15}$$

Einsteinium was detected in the remnants of thermonuclear devices detonated at Eniwetok Island in the Pacific. It was formed from multiple neutron capture of uranium-238 accompanied by six virtually concomitant beta decays to give californium-253; one further beta decays produces einsteinium-253. Fermium, like einsteinium, was discovered as a product of thermonuclear testing. Bombardment of einsteinium-253 with alpha particles yields mendelevium:

$$^{253}_{99}\text{Es} + ^{4}_{2}\text{He} \longrightarrow ^{256}_{101}\text{Md} + ^{1}_{0}\text{n} \tag{11.16}$$

The final two elements in the series, nobelium and lawrencium, have been synthesized by bombarding heavy elements with moderately heavy ions. So, we see that nobelium can be prepared from curium by bombardment with carbon ions. Similarly, lawrencium is prepared by bombardment of californium with boron ions.

In common with many of the early actinoids, these latter elements exist chiefly in the Ac(III) oxidation state. They can be separated by ion-exchange on a strong cation exchange resin (e.g., Doxlex-50) and eluting the various elements with an appropriate chelating solvent such as ammonium α-hydroxyisobutyrate.

Chapter Summary

- The f-block elements are known as the lanthanoids (Period 6) and the actinoids (Period 7). The lanthanoids are relatively abundant in the Earth's crust, while the radioactive nature of the actinoids limits the abundance of certain isotopes.
- The majority of the lanthanoids exist in the Ln(III) oxidation state, with the exception of cerium which is dominated by the Ce(IV) state. Lanthanoids exhibit fairly typical metal chemistry through formation of oxides, hydroxides and halides.
- The actinoids are a series of fifteen radioactive elements formed by neutron or heavy atom bombardment. Uranium and plutonium are the most commercially valuable actinoids.

Review Questions

(1) Define the lanthanioid contraction and suggest how lanthanioids can be separated from their ores.

(2) Identify the isotopes A, B, and C in the sequence below:

$$A \xrightarrow{-\beta^-} B \xrightarrow{(n,\gamma)} {}^{242}Am \xrightarrow{-\beta^-} C$$

(3) Using Nd as an example, explain why the elements in Period 6 favor the Ln(III) oxidation state.

(4) Outline the general reactions of an aqueous lanthanoid with carbon dioxide and fluoride ion.

(5) Provide an explanation for the linear geometry of the $[UO_2]^{2+}$ cation in the solid state.

(6) What is the systematic name for $[UO_2Cl_4]^{2-}$?

(7) Suggest plausible equations for the conversion of uranium dioxide to ammonium diuranate.

(8) Provide an equilibrium expression for the hydrolysis of PuO_2^+ and comment on the stability of the product at low pH.

(9) What is the maximum volume of uranium hexafluoride that can be prepared from 1 kg of uranium dioxide at rtp?

(10) Develop a scheme to show the production of americium dioxide from spent nuclear fuel.

Chapter 12

Nuclear Power

"Science is not an abstraction; but as a product of human endeavor it
is inseparably bound up in its development with the personalities and
fortunes of those who dedicate themselves to it."

E. Fischer

Meeting the world energy demands is a key challenge for 21st century
scientists. Since the discovery of nuclear fission, the potential of nuclear
power has been recognized and the large world powers have well-developed
nuclear power programs. On completion of this chapter and the associated
questions you should:

- Be able to identify the chemical stages in the preparation and enrichment
  of nuclear fuels.
- Be able to identify the common features of a nuclear power reactor,
  and the key differences between pressurized water reactors (PWRs) and
  boiling water reactors (BWRs).
- Understand recent developments in nuclear power technology, and the
  role of public opinion in shaping nuclear policy.

12.1 Introduction

Key Point: The modern nuclear power industry stemmed from the
Manhattan Project.

The discovery of artificial nuclear fission in 1938 by Hahn, Strassman,
and Meitner led many scientists to believe that it should be possible to

use the neutrons released during fission to sustain a nuclear chain reaction, producing large amounts of energy. The first successful attempt came on December 2, 1942 when the "Chicago pile" achieved critical mass (the point at which the nuclear reaction is self-sustaining). This project was initially under the supervision of Enrico Fermi as part of the Manhattan Project, but after the conclusion of the Second World War, the application of nuclear reactors for civilian use (electrical power generation) became clear. The first such plant was built in Russia and started to generate electrical power in 1954. Since then, many civilian nuclear reactors have been constructed around the world, with the USA and France generating the largest amounts of electricity using nuclear power. Nuclear power is also used in marine propulsion, with reactors being deployed in many of the world's naval vessels (mostly submarines). Russia has a fleet of six nuclear powered icebreakers and is currently developing two floating nuclear power stations to meet energy demands in remote areas.

12.2 Nuclear Fuel

Key Point: Most modern nuclear reactors required uranium fuels enriched in the uranium-235 isotope.

In common with traditional power generation using fossil fuels, the nuclear power industry requires a fuel source. The main fissionable fuel used in modern nuclear power reactors is ***uranium dioxide***, UO_2, which is obtained from a number of uranium mines across the world. In its natural form, uranium typically exists in the tetravalent state and is found in ores containing a variety of other metal oxides, which could potentially poison the fission process. These must be removed before further use.

Initially, the ore is mined, crushed, and treated with sulfuric acid which dissolves the uranium content and produces a uranyl sulfate complex:

$$UO_3 + 2H^+ \rightarrow UO_2^{2+} + H_2O$$

$$UO_2^{2+} + 3SO_4^{2-} \rightarrow [UO_2(SO_4)_3]^{4-}$$

The uranyl sulfate complex is known as ***the feed***. This complex is extracted into an organic solvent, usually a tertiary amine such as octadecylamine dissolved in kerosene. In this procedure, the solvent is first converted to a sulfate and then used to extract the uranyl complex:

$$2R_3N + H_2SO_4 \rightarrow (R_3NH)_2SO_4$$

$$2(R_3NH)_2SO_4 + [UO_2(SO_4)_3]^{4-} \rightarrow (R_3NH)_4UO_2(SO_4)_3 + 2SO_4^{2-}$$

As the uranyl complex is relatively stable, cationic and anionic contaminants can be removed by adjusting the pH with sulfuric acid and gaseous ammonia. The complex is then recovered by treatment with ammonium sulfate:

$$(R_3NH)_4UO_2(SO_4)_3 + 2(NH_4)_2(SO_4)$$
$$\rightarrow 4R_3N + (NH_4)_4UO_2(SO_4)_3 + 2H_2SO_4$$

Finally, ammonium diuranate is precipitated from the solution by bubbling through ammonia:

$$2[UO_2(SO_4)_3]^{4-} + 2NH_3 \rightarrow 4R_3N + (NH_4)_2U_2O_7 + 4SO_4^{2-}$$

Ammonium diuranate is roasted to give U_3O_8 which is often referred to as *yellowcake*. The purity of yellowcake uranium is such that it can be handled in drums without radiation protection suits. For this reason, it is the preferred form of uranium for shipping.

The naturally occurring uranium dioxide which is used to produce yellowcake contains approximately 0.7% uranium-235, the fissionable isotope required for a nuclear chain reaction. While some nuclear reactors can operate with this level of uranium-235, most are designed to utilize uranium enriched to around 2% uranium-235. The enrichment process begins by dissolving the yellowcake in nitric acid producing uranyl nitrate, $UO_2(NO_3)_2 \cdot 6H_2O$. The uranium content is then recovered by solvent extraction using tributyl phosphate in kerosene and the uranium extract calcined to produce UO_3. This is reduced by hydrogen to give uranium dioxide which is then reacted with gaseous hydrogen fluoride to give uranium tetrafluoride:

$$UO_3 + H_2 \rightarrow UO_2 + H_2O$$
$$UO_2 + 4HF \rightarrow UF_4 + 2H_2O$$

The uranium tetrafluoride is further fluoridated to give *uranium hexafluoride* ("hex"):

$$UF_4 + F_2 \rightarrow UF_6$$

which is a grey, volatile crystalline solid at room temperature. It has a relatively simple phase diagram (Figure 12.1) which shows that it sublimes at 56.5°C and has a triple point at 64°C. The density of UF_6 at room temperature is around 5.1 g/cm^3; however, a large decrease in density occurs during the change solid $\rightarrow$ liquid which is accompanied

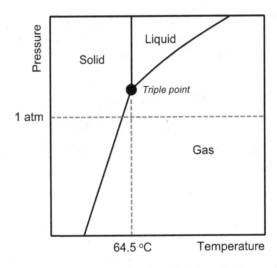

Figure 12.1 Phase diagram for uranium hexafluoride. The solid is volatile and sublimes at room temperature.

by a large increase in volume. This must be taken into account when storing UF_6.

Enrichment of uranium hexafluoride in the uranium-235 isotope is almost always performed by *gas centrifugation*. Other technologies such as gaseous diffusion and laser processing have been used, but for large-scale production the centrifuge method remains the most popular. The concept is relatively simple: the uranium hexafluoride is piped into an evacuated cylindrical centrifuge tube (about $20 \times 300\,cm^2$) and when the tube is spun at high speeds, the heavier uranium-238 moves toward the outer edge of the tube, separating it from the uranium-235. The heavier $^{238}UF_6$ is then transferred into another centrifuge tube and the process repeated successively until an enrichment of about 3% is obtained.

Enriched uranium hexafluoride is converted to uranium dioxide using the so-called wet or dry process. In the former, UF_6 is mixed with water, producing a slurry of uranyl fluoride. This is then reacted with ammonia to produce ammonium diuranate. This is recovered and heated in a reducing atmosphere, producing uranium dioxide. In the dry process, UF_6 is reacted with steam, producing uranyl fluoride, which is then reduced by hydrogen to produce the uranium dioxide:

$$UF_6 + 2H_2O \rightarrow UO_2F_2 + 4HF$$
$$UO_2F_2 + H_2 \rightarrow UO_2 + HF$$

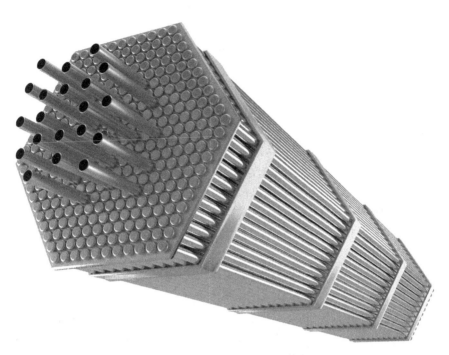

Figure 12.2 Nuclear fuel assembly. Tubes containing uranium are typically arranged into a hexagonal fuel assembly which forms the basis of the nuclear reactor core.

Regardless of which process is used, the uranium dioxide typically needs to be homogenized before being compressed into pellets. Uranium dioxide pellets are held in rods, forming *fuel rods*. These are in turn arranged into fuel assemblies which are the main site of fission in the reactor core (Figure 12.2).

12.3 Nuclear Power Reactors

Key Point: Current nuclear power reactors use fission of enriched uranium to generate steam which drives turbines for electrical power.

Around 11% of the world's total electricity supply comes from nuclear power with over four hundred nuclear power stations in operation around the world. Although nuclear reactor designs vary, there are a number of common features which are essential for the operation of a nuclear power facility. Firstly, there needs to be a supply of nuclear fuel, which as discussed in the previous section consists of pellets of uranium dioxide packed to form fuel rods. These fuel rods are held within the core of the reactor which is filled with a neutron *moderator*, usually (light) water or, more rarely,

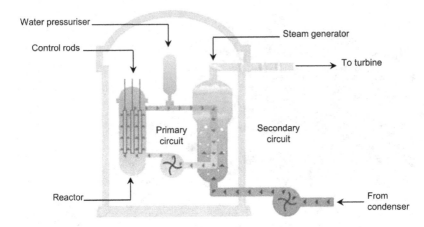

Figure 12.3 Pressurized water reactor. The general assembly of a PWR includes a water pressurizer which increases the neutron moderation efficiency. The steam drives a turbine and is then condensed in large cooling towers before returning to the steam generator.

heavy water. The moderator slows down neutrons produced in the decay reaction. A number of **control rods**, usually made of hafnium or boron, are inserted or withdrawn from the core to control the rate of the nuclear reaction. The **coolant** flows through the reactor and carries away the heat produced during the decay process. In some reactors, the water used to moderate the reaction also serves as the coolant. The entire assembly is contained with a **pressure vessel** which is in turn surrounded by a concrete and steel structure to prevent the release of ionizing radiation.

The design engineering of a nuclear power reactor is based on the propulsion systems for early submarines and naval warships. A process, in this case nuclear fission, generates heat which is used to create steam, which then drives a turbine to generate the electricity. Most reactors currently in use are the **pressurized water reactors** (PWR) (Figure 12.3) which require enriched uranium fuel. PWRs use light water as both the moderator and coolant which flow at high pressure through a primary cooling circuit. A secondary circuit utilizes the steam for generation of electricity. The reactor core of a PWR contains around one hundred tons of uranium held in vertical fuel rods. The average temperature of the reactor core is 325°C, well above the normal boiling point of water. To prevent boiling, the pressure vessel maintains the core at 150 atm using a pressurizer unit. When the water flows into the secondary circuit, it experiences lower pressure and starts to boil, generating steam, and driving the turbines.

The *boiling water reactor* (BWR) is similar in principle to the PWR, except that is consists of a single circuit under lower temperature and pressure (285°C, 75 atm) and it utilizes less enriched uranium (140 tons). Steam collects in the head of the reactor core and so the neutron moderation is less efficient. Also, since the moderator/cooling water is in direct contact with the core, and this same water is used to drive the turbine, there must be greater radiation shielding around the turbine assembly.

The *pressurized heavy water reactor* (PHWR) was originally developed in Canada (CANDU) and is also now used in India. These reactors use natural uranium fuels and therefore need a more efficient moderator, in this case heavy water. The reactor consists of several hundred-pressure tubes, which contain the fuel rods. The moderator/coolant flows through the pressure tubes, flowing toward a steam generator linked to a turbine system. An advantage of the PHWR is that it can be operated continuously; it does not need to be shut down for refueling. In addition, it can also use a greater variety of fuel types — for example, depleted uranium.

There are a number of other reaction technologies in use or in development. In the UK, *gas-cooled reactors* such as the now obsolete Magnox reactor were common. These utilized graphite moderators and pressurized carbon dioxide coolant. The fuel rods were composed of natural uranium which was clad with a magnesium-aluminum alloy. These have now been replaced by second generation advanced gas-cooled reactors which use enriched uranium fuels. A summary of the reactor types in use around the world is given in Table 12.1.

Table 12.1 Commerical Nuclear Power Plants Around the World

Reactor Type*	Main Countries	Total Number	Fuel	Coolant	Moderator
PWR	US, France, Japan, Russia, Chinna	277	Enriched UO_2	H_2O	H_2O
BWR	US, Japan, Sweden	80	Enriched UO_2	H_2O	H_2O
PHWR	Canada, India	49	Natural UO_2	D_2O	D_2O
GCR	UK	15	Enriched UO_2	CO_2	Graphite
LWGR	Russia	15	Enriched UO_2	H_2O	Graphite
FNR	Russia	2	PuO_2 and UO_2	Liquid Na	None

*PWR, pressurized water reactor; BWR, boiling water reactor; PHWR, pressurized heavy water reactor; GCR, gas-cooled reactor; LWGR, light water graphite reactor; FNR, fast neutron reactor.

12.4 Nuclear Fuel Reprocessing

Key Point: Used nuclear fuels are reprocessed and the uranium/plutonium content extracted and recycled. This forms a key element of the nuclear fuel cycle.

Spent nuclear fuels pose a very obvious risk to the environment; aside from the radiological effects, depleted uranium has military applications, so particular care must be taken when transporting and storing nuclear waste. Due to the commercial value of some of the products of uranium decay, chiefly plutonium, spent nuclear fuels are typically reprocessed to extract these valuable materials. The reprocessing of spent nuclear fuel also represents a smaller environmental burden and in recent years there has been renewed interest in recovering long-lived actinides. Moreover, processing used fuels to recover fertile uranium is a worthwhile undertaking; about 96% of used nuclear fuel is uranium-235 which can be recycled and used as fresh fuel. This obviously reduces the need for extensive uranium mining and preserves what is a finite fuel source.

Most commercial reprocessing plants employ the *PUREX* (plutonium uranium extraction) process, which is very similar to the uranium extraction process described in Section 12.3. The fuel elements are dissolved in concentrated nitric acid and the uranium/plutonium content separated by solvent extraction. Recovered uranium is sent to the conversion plant for enrichment; recovered plutonium is used for production of *mixed oxide fuel* (MOX). Often a small quantity of neptunium is also recovered and this can be used for production of plutonium-238 for use in the power plant of spacecraft. The remaining decay products are held in the liquid waste from the extraction process. This waste is highly radioactive and generates large amounts of heat. In the 1960s, a *vitrification* process was developed for disposal of high-level radioactive waste. This involves evaporating off the solvent from liquid waste and mixing the solids with glass-forming materials. This is then melted and poured in metal canisters which are sealed; these are then stored in a secure underground location which is geologically stable.

12.5 The Future for Nuclear Power

Key Point: New nuclear power technologies are becoming an increasingly important way to meet the world's energy demands.

At present, sixteen countries depend on nuclear power for at least 25% of their electricity, with France having the highest dependency at 75%. Across the world, governments are planning to construct new nuclear power facilities, initially to replace ageing reactors, but to also increase capacity. China has invested heavily in nuclear technologies and has completed construction of 28 new nuclear reactors in the past 14 years with a further 24 in the planning stages. India has an emergent and ambitious nuclear power program which seeks to employ *molten salt reactors* using thorium as fuel.[a]

Modern molten salt reactors are being developed in which the fertile and/or fissile fuel is dissolved in a molten lithium fluoride coolant, forming a *fuel salt*. As this solution remains molten at standard pressure, there are significant savings to be made in terms of energy input. The basic design involves heating the lithium fluoride and uranium-233 fluoride salts to around 700°C in a reactor core with a graphite moderator. As fluoride salts have a low vapor pressure even at elevated temperatures, they can carry more heat than the same volume of water, and so reactors can be reduced in size. The heat is transferred to a secondary circuit, creating steam and eventually electricity. As the fission products will be dissolved in the molten salt, they can be removed periodically by an online reprocessing loop. In India, where the use of thorium-232 is intended, liquid fluoride thorium reactors are being designed.

A large area of research is focused on the development of nuclear fusion reactors, such as the *tokamak reactor* which was first developed in Russia in the 1970s. The principle is essentially the same as that employed by stars which fuse together different isotopes of hydrogen to create helium plus large amounts of energy. The challenge is to create high enough temperatures to initiate fusion, while containing the reaction to maintain sufficient density to allow the reaction to proceed. Theoretically, once the ignition temperature has been reached, the reaction is self-sustaining, only needed addition of fresh fuel.

Current models are based on fusing deuterium and tritium atoms, liberating neutrons which are absorbed by lithium. The kinetic energy absorbed by a blanket of lithium surrounding the core is converted to

---

[a]India has large thorium oxide deposits around its south and eastern coastlines, believed to total some 846000 tons.

heat, which can then be used to produce electricity as in conventional nuclear power reactors. The neutron–lithium reaction produces tritium, which can fuel the reaction, and helium. While generating a high enough temperature for the deuterium–tritium reaction is possible, the issue of containment proved more problematic. At present, two approaches are used: magnetic confinement and internal confinement. In the former, two overlapping magnetic fields create a helical confinement field which repels the charged ions in the reactor core, maintaining sufficient density. In the other approach, a laser is focused on a small pellet of deuterium–tritium fuel, which causes an implosion, increasing the density of the inner fuel layers. The heat released then causes fusion of the surrounding fuel, liberating heat which is harnessed for production of electricity as before.

Nuclear power is a massive subject which could easily fill several textbooks on its own. A very brief overview has been presented here to give an indication of the increasing importance of nuclear technologies. Interested readers can find up-to-date information on the World Nuclear Association's website: http://www.world-nuclear.org/

Chapter Summary

- Uranium dioxide is the raw material for production of nuclear fuel. The uranium is extracted in a series of solvent extraction steps and converted to uranium hexafluoride. This can be enriched in uranium-235 to produce enriched uranium fuel rods.
- Nuclear power reactors harness the energy from a controlled fission reaction to form steam, which in turn is used to drive a turbine, generating electricity. Most modern reactors use a pressurized water design, which requires enriched uranium fuel.
- The products of the fission process constitute "nuclear waste" and must be carefully managed. This initially involves recovery of valuable material, chiefly plutonium and uranium, through the PUREX process. High-level waste is vitrified and stored deep underground.
- Modern advancements in nuclear technology include the development of molten salt reactors and fusion reactors. The latter have huge potential for creation of clean energy.

Review Questions

(1) Show that uranium in UO_2 is in the tetravalent state.

(2) Suggest why the pH of the tertiary amine–uranyl complex is adjusted using gaseous ammonia.

(3) Why is yellowcake uranium safe to handle in metal drums without wearing radiation protection suits?

(4) In terms of isotopologues, suggest why uranium as uranium hexafluoride is an ideal compound for enrichment by gas centrifugation.

(5) Suggest reasonable equations for the conversion of uranium hexafluoride to uranium dioxide by the wet process.

(6) The water used to moderate the reaction within a nuclear reactor often contains boric acid. Suggest why this is an advantage.

(7) Why is neutron moderation less efficient in the boiling water reactor?

(8) In the PUREX process, spent fuel elements are first dissolved in concentrated nitric acid. Why is nitric acid the best choice for this process?

(9) The thorium fuel cycle is based on neutron capture by thorium-232, initially forming thorium-233; two beta decay steps produce uranium-233. Provide a radiochemical equation for this process.

(10) Fusion of deuterium and tritium has been shown to be possible; however, the artificial fusion of two deuterium isotopes has yet to be shown. Suggest why this is the case.

Appendix A

Review of Mathematics

There are a many great mathematics textbooks, some of which are specifically aimed at science students. A few particularly useful examples are:

1. Steiner, E. 2008. *The Chemistry Maths Book*. Oxford University Press, UK.
2. Monk, P. 2010. *Maths for Chemistry*. Oxford University Press, UK.
3. Cornish-Bowden, A. 1999. *Basic Mathematics for Biochemists*. Oxford University Press, UK.

In addition, the Maths Tutor website, www.mathstutor.ac.uk, provides excellent guides on many aspects of mathematics.

Rules of Differentiation

The association between variables in an equation could involve powers, logarithms, fractions or trigonometric functions. When differentiation is used to determine the rate of change of one of these variables with the other, different rules are used depending on how the variables are related. Some of the basic rules are shown below.

1. $y = ax^n,\quad \dfrac{dy}{dx} = (n \times a)\, x^{(n-1)}$

4. $y = \ln x,\quad \dfrac{dy}{dx} = \dfrac{1}{x}$

2. $y = e^{ax},\quad \dfrac{dy}{dx} = a\, e^{ax}$

5. $y = \sin ax,\quad \dfrac{dy}{dx} = a \cos ax$

3. $y = \cos ax,\quad \dfrac{dy}{dx} = -a \sin ax$

6. $y = \tan ax,\quad \dfrac{dy}{dx} = a \sec^2 ax$

Rules of Integration

Integration can be considered as the inverse of differentiation and is often used to solve differential equations. There are extensive tables of standard integral available online; a brief summary is shown below. Note that the constant of integration is omitted for simplicity.

1. $\int x^n \, dx = \dfrac{x^{n+1}}{n+1}$ $(n \neq -1)$ 5. $\int \sin ax \, dx = -\dfrac{1}{a} \cos ax$

2. $\int \ln ax \, dx = x \ln ax - x$ 6. $\int \cos ax \, dx = \dfrac{1}{a} \sin ax$

3. $\int e^{ax} \, dx = \dfrac{1}{a} e^{ax}$ 7. $\int \tan ax \, dx = -\dfrac{1}{a} \ln \cos ax$

4. $\int \dfrac{1}{x} dx = \ln x$ 8. $\int \dfrac{1}{\sqrt{(a^2 - x^2)}} dx = \sin^{-1}\left(\dfrac{x}{a}\right)$

The Harmonic Oscillator

Integration of the equation presented in Chapter 1 for the harmonic oscillator (Eq. (1.22)) involves more extensive use of the rules of calculus as well as some basic algebra. It is a useful exercise for those more familiar with mathematics. The approach is as follows:

(1) We are dealing with a function of a function; therefore, we can simplify the second-order differential using the chain rule of calculus:

$$\frac{d^2x}{dt^2} = \frac{dv}{dt} = \frac{dv}{dx} \cdot \frac{dx}{dt} = \frac{dv}{dx} v, \quad \therefore \frac{dv}{dx} v = -\left(\frac{\kappa}{m}\right) x$$

(2) Integrating the previous result (remembering that $-\kappa/m$ is a constant):

$$\int \frac{dv}{dx} v = -\left(\frac{\kappa}{m}\right) \int x, \quad \Rightarrow \frac{v^2}{2} = -\left(\frac{\kappa}{m}\right) \cdot \frac{x^2}{2} + C$$

(3) To evaluate the constant of integration, we set the condition $v = 0$ which means that x will be equal to the amplitude of the wave, a:

$$0 = -\left(\frac{\kappa}{m}\right) \cdot \frac{a^2}{2} + C, \quad \therefore C = \left(\frac{\kappa}{m}\right) \cdot \frac{a^2}{2}$$

(4) Substituting in the expression for C and simplifying:

$$\frac{v^2}{2} = -\left(\frac{\kappa}{m}\right) \cdot \frac{x^2}{2} + \left(\frac{\kappa}{m}\right) \cdot \frac{a^2}{2} \xrightarrow{\times 2} v^2 = \left(\frac{\kappa}{m}\right)\left(a^2 - x^2\right)$$

(5) Taking square roots and restating v as a differential:

$$v = \frac{dx}{dt} = \sqrt{\left(\frac{\kappa}{m}\right)\left(a^2 - x^2\right)}$$

(6) Separating the variables to make the integration more obvious:

$$\frac{1}{dt} = \frac{\sqrt{\left(\frac{\kappa}{m}\right)\left(a^2 - x^2\right)}}{dx} \Rightarrow \sqrt{\left(\frac{\kappa}{m}\right)} \cdot dt = \frac{dx}{\sqrt{\left(a^2 - x^2\right)}}$$

(7) Completing the integration:

$$\int \frac{dx}{\sqrt{\left(a^2 - x^2\right)}} = \sqrt{\left(\frac{\kappa}{m}\right)} \cdot \int dt \Rightarrow \sin^{-1}\left(\frac{x}{a}\right) = \sqrt{\left(\frac{\kappa}{m}\right)} \cdot t + C$$

(8) Simplifying and rearranging gives the basic equation for a sine wave, verifying that the harmonic oscillator will have a sinusoidal form:

$$\text{Let } \sqrt{\frac{\kappa}{m}} = \omega \quad \text{and} \quad C = \phi, \quad \therefore \quad x = a\sin\left(\omega t + \phi\right)$$

Appendix B

Solutions to Numerical Questions

Chapter 1

Q1. (i) $v = 2 \times 0^2 - 7 \times 0 + 3 = 3$; (ii) $a = 4t - 7 = 0$, $\therefore$ $t = 1.75$;

(iii) $s = \dfrac{2}{3}t^3 - \dfrac{7}{2}t^2 + 3t + c$, $\therefore$ $s = \dfrac{17}{24}$

Q2. (i) $p = 70 \times 5.2 = 364\,\text{kg m/s}$; (ii) $F = 364/0.832 = 437.5\,\text{N}$

Q3. (i) $K = \dfrac{1}{2} \times 900 \times 26.8^2 = 323208\,\text{J}$;

(ii) $K = \dfrac{1}{2} \times 900 \times 13.4^2 = 80802\,\text{J}$

Q4. $L = \left(7.35 \times 10^{22}\right) \times \left[\left(2\pi \times 3.84 \times 10^8\right) / \left(31 \times 24 \times 60 \times 60\right)\right]$
$\times \left(3.84 \times 10^8\right) = 2.54 \times 10^{34}\,\text{kg m}^2/\text{s}$

Q5. $\psi\,(6, 1) = 0.05 \times \sin\left[\dfrac{2\pi}{4}\,(6 - 5 \times 1)\right] = 0.05\,\text{m}$

Q6. $\lambda = \dfrac{2.99 \times 10^8}{3 \times 10^3} = 9.97 \times 10^4\,\text{m} = 99.7\,\text{km}$

Q7. $E = \left(2 \times 1.67 \times 10^{-27}\right) \times \left(2.99 \times 10^8\right)^2 = 2.99 \times 10^{-10}\,\text{J}$

Q8. (i) $F = \dfrac{1 \times 1}{4\pi \times 8.85 \times 10^{-12} \times 0.25^2} = 1.44 \times 10^{11}\,\text{N}$;

(ii) $F = \dfrac{1 \times 1}{4\pi \times 8.85 \times 10^{-12} \times 0.50^2} = 3.60 \times 10^{10}\,\text{N}$;

(iii) $F = \dfrac{1 \times 1}{4\pi \times 8.85 \times 10^{-12} \times 1^2} = 8.99 \times 10^9\,\text{N}$

Chapter 2

Q2. $\lambda = \dfrac{\left(6.62 \times 10^{-34}\right) \times \left(2.99 \times 10^{8}\right)}{\left(4.73 \times 1.60 \times 10^{-19}\right)} = 2.61 \times 10^{-7}\,\mathrm{m}$

Q3. $\lambda = \left[\left(1.097 \times 10^{7}\right) \times \left(\dfrac{1}{4} - \dfrac{1}{9}\right)\right]^{-1} = 6.56 \times 10^{-7}\,\mathrm{m}$

Q5. $\lambda = \dfrac{6.6 \times 10^{-34}}{70 \times 3.1 \times 0.447} = 6.8 \times 10^{-36}\,\mathrm{m}$

Chapter 3

Q3. B.E. $= \left[92\left(1.007825 \times 931.48\right)\right] + \left[143\left(1.008665 \times 931.48\right)\right]$

$\qquad - \left(235 \times 931.48\right) = 1780\,\mathrm{MeV} = \dfrac{1780}{235} = 7.59\,\mathrm{MeV/nucleon}$

Q4. $R(\mathrm{Li}) = 1.2 \times \sqrt[3]{7} = 2.3\,\mathrm{fm}; \quad R(Pu) = 1.2 \times \sqrt[3]{239} = 7.4\,\mathrm{fm}$

Q5. $\rho = \dfrac{27 \times 1.66 \times 10^{-27}}{4/3 \times \pi \times \left(1.2 \times 10^{-15} \times \sqrt[3]{27}\right)^{3}} = \dfrac{4.48 \times 10^{-26}}{1.96 \times 10^{-43}}$

$\qquad = 2.29 \times 10^{17}\,\mathrm{kg/m^{3}}$

Q8. $\rho = \dfrac{1.67 \times 10^{-24}}{4/3 \times \pi \times \left(1 \times 10^{-13}\right)^{3}} = 4.0 \times 10^{14}\,\mathrm{g/cm^{3}}$

Q10. B.E.(Cu) $= \left(15.7 \times 64\right) - \left(17.8 \times 64^{2/3}\right)$

$\qquad - \left[0.71\dfrac{29\left(29 - 1\right)}{64^{1/3}}\right] - \left[23.6\dfrac{\left(35 - 29\right)^{2}}{64}\right] - 11.8$

$\qquad = 1004.8 - 284.8 - 144.1 - 13.3 - 11.8 = 563.5\,\mathrm{MeV}$

$\qquad = \dfrac{563.5}{64} = 8.8\,\mathrm{MeV/nucleon}$

B.E.(Zn) $= \left(15.7 \times 64\right) - \left(17.8 \times 64^{2/3}\right)$

$\qquad - \left[0.71\dfrac{30\left(30 - 1\right)}{64^{1/3}}\right] - \left[23.6\dfrac{\left(35 - 30\right)^{2}}{64}\right] + 11.8$

$\qquad = 1004.8 - 284.8 - 154.4 - 9.2 + 11.8 = 568.2\,\mathrm{MeV}$

$\qquad = \dfrac{568.2}{64} = 8.8\,\mathrm{MeV/nucleon}$

Chapter 4

Q4. $Q_\alpha = -931.502 \, (220.0113940 - 216.0019150 - 4.002603)$

$$= 6.405 \, \text{MeV}$$

$$K_\alpha = \frac{6.405 \times 216.0019150}{220.0113940} = 6.29 \, \text{MeV}$$

Q6. $Q_{\beta+} = -931.502 \times [13.003355 + (2 \times 0.000549) - 13.00573861]$

$$= 1.2 \, \text{MeV}$$

Q8. $Q = -931.502 \times (1.008665 - 1.007825) = 0.8 \, \text{MeV}$

Q10. $\lambda = \dfrac{6.63 \times 10^{-34} \times 2.99 \times 10^{8}}{1.33 \times 1.602 \times 10^{-13}} = 9.3 \times 10^{-13} \, \text{m}$

Chapter 5

Q1. $N = 37 \times 10^{6} / \left[\ln 2 / (8.02 \times 24 \times 60 \times 60)\right] = 3.7 \times 10^{13} \, \text{atoms}$

$n = 3.7 \times 10^{13} / 6.02 \times 10^{23} = 6.146 \times 10^{-11} \, \text{mol m}$

$= 6.146 \times 10^{-11} \times 131 = 8 \times 10^{-9} \, \text{g}$

Q2. $\alpha/s = \dfrac{114}{4} \times 60 \times 60 = 102600; \quad 5 \times 0.2\% = 0.01 \, \text{g};$

$\dfrac{0.01}{152} \times 6.02 \times 10^{23} = 3.96 \times 10^{19} \, \text{atoms} \; t_{1/2} = \ln 2 / 3.96 \times 10^{19}$

$= 3.469 \times 10^{21}, s = 1.10 \times 10^{14} \, \text{years}$

Q3. $m = 75 \times 0.35\% \times 0.012\% \times 10^{3} = 0.0315 \, \text{g}$

$\Rightarrow \; (0.0315/40) \times 6.02 \times 10^{23} = 4.74 \times 10^{20} \, \text{atoms} A$

$= \left[\ln 2 / \left(1.3 \times 10^{9} \times 365 \times 24 \times 60\right)\right] \times 4.74 \times 10^{20} = 480844.9 \, \text{dpm}$

Q4. $N = 2 \times 10^{-3} \times 6.02 \times 10^{23}$

$= 1.204 \times 10^{21} \, \text{atoms} \; (2 \, \text{mmol}^{32}\text{P per molecule})$

$A = \left[\ln 2 / (14.3 \times 24 \times 60 \times 60)\right] \times 1.204 \times 10^{21}$

$= 6.75 \times 10^{5} \, \text{GBq/mmol}$

Q5. $\lambda = \ln 2 / (5730 \times 365 \times 24 \times 60 \times 60) = 3.84 \times 10^{-12} \, \text{s}^{-1}$

$A_0 = \left(3.84 \times 10^{-12}\right) \times \left(8.50 \times 10^{-6}\right) \times \left(6.02 \times 10^{23}\right) = 19649280 \, \text{Bq}$

$A_t = \left(3.84 \times 10^{-12}\right) \times \left(0.80 \times 10^{-6}\right) \times \left(6.02 \times 10^{23}\right) = 1849344 \, \text{Bq}$

$t = -\ln (1849344/19649280) / 3.84 \times 10^{-12} = 6.16 \times 10^{11} \, \text{s}$

$= 1.95 \times 10^{4} \, \text{years}$

Q6. $\lambda = \ln 2 / (4.5 \times 24 \times 60 \times 60) = 1.78 \times 10^{-6} \, \text{s}^{-1} N_t$

$= \dfrac{5 \times 10^{-6}}{47} \times 6.02 \times 10^{23} = 6.4 \times 10^{16} \, \text{atoms}$

$$N_0 = \frac{6.4 \times 10^{16}}{\exp\left(-1.78 \times 10^{-6} \times 3.88 \times 10^5\right)} = 1.28 \times 10^{17} \, \text{atoms}$$

$$m = \left(1.28 \times 10^{17}/6.02 \times 10^{23}\right) \times 74 = 1.5 \times 10^{-5} \, \text{g}$$

Q7. $0.1 \times 5000 = 500 \, \text{cps}; \quad 500/48 \approx 10 \, \text{mL}$

Q8. $A_2/A_1 = 0.2449/\left(0.2449 - 0.0087\right) = 1.0368$

$$A_2 = 11.1 \times 1.0368 \times \{1 - [\exp - (0.2449 - 0.0087) \times 6]\}$$

$$= 8.7 \, \text{GBq}$$

Q9. $N = 2.0 \times 10^{-6}/\left(\ln 2/0.07 \times 10^{-15}\right) = 2.02 \times 10^{-22} \, \text{atoms}$

$A_0 = 9.9 \times 10^{15} \times 2.02 \times 10^{-22} = 2 \times 10^{-6} \, \text{Bq}; \quad A_t = 2 \times 10^{-9} \, \text{Bq}$

$t = \ln(2 \times 10^{-9}/2 \times 10^{-6})/9.9 \times 10^{15} = 6.9 \times 10^{-16} \, \text{s}$

Q10. $N_0 = (25/210) \times 6.02 \times 10^{23} = 7.17 \times 10^{22} \, \text{atoms}$

$N_t = 7.17 \times 10^{22}/2 = 3.58 \times 10^{22} \, \text{atoms} = 0.0596 \, \text{mol}$

$V = 0.0596 \times 24 = 1.43 \, \text{dm}^3$

Chapter 6

Q2. $Q = (14.00307 + 4.00260 - 16.99913 + 1.007825) \times \left(2.99 \times 10^8\right)^2$

$$= 1.8 \times 10^{17} \, \text{J}$$

Q5. $\lambda = \ln 2/(5.27 \times 365 \times 24 \times 60 \times 60) = 4.0 \times 10^{-9} \, \text{s}^{-1}$

$n = (0.06/59) \times 6.02 \times 10^{23} = 6.12 \times 10^{20} \, \text{nuclei/mg}$

$t_{\text{irr}} = 37 \times 10^6/\left[\left(5 \times 10^{13}\right) \times \left(37 \times 10^{-24}\right) \times \left(6.12 \times 10^{20}\right)\right.$
$\left. \times \left(4.0 \times 10^{-9}\right)\right] = 2.2 \, \text{h}$

Q6. $1 \, \mu\text{A} = 3.1 \times 10^{12} \, \alpha/\text{s}; \quad \lambda = \ln 2/(270 \times 24 \times 60) = 1.78 \times 10^{-6} \, \text{min}^{-1}$

$n = (0.01/55) \times 6.02 \times 10^{23} = 1.09 \times 10^{20}$

$A = 3.1 \times 10^{12} \times 200 \times 10^{-27} \times 1.09 \times 10^{20} \times 1.78 \times 10^{-6} \times 3600$

$$= 4.3 \times 10^5 \, \text{min}^{-1}$$

Q7. $n = (0.02/65) \times 6.02 \times 10^{23} = 1.85 \times 10^{20}$

$\sigma = \left(2.0 \times 10^5/1.0 \times 10^7\right)/1.85 \times 10^{20} = 1.08 \times 10^{-22} b$

Q8. $K.E. = 9.0 \times 10^9 \times \left(1.6 \times 10^{-19}\right)^2/2.0 \times 10^{-15} = 1.15 \times 10^{-13} \, \text{J}$

$T = 2 \times 5.75 \times 10^{-14}/3 \times 1.38 \times 10^{-23} = 2.78 \times 10^9 \, \text{K}$

Q9. Overall reaction: $4\,{}_1^1\mathrm{H} \rightarrow {}_2^4\mathrm{He}$

$\therefore Q = 931.478\,(4 \times 1.007825 - 4.002602) = 26.7\,\mathrm{MeV}$

Q10. $Q = 931.478[(235.043915 + 1.008665) - (140.9139 + 91.8973 + 3 \times 1.008665)] = 200\,\mathrm{MeV}$

Chapter 7

Q4. $\dot{D}_0 = (5.20 \times 10^{-8} \times 833)/3.0^2 = 4.81 \times 10^{-6}\,\mathrm{Gy/h}$

Q5. $\dot{D}_0 = \ln\left(9.6 \times 10^{-5}\right) = \ln\left(2.85 \times 10^{-2}\right) \times \left(-\dfrac{(\ln 2)\,x}{1.2}\right)$

$$x = -\frac{\left[\ln\left(2.86 \times 10^{-9}\right) - \ln\left(2.85 \times 10^{-2}\right)\right] \times 1.2}{\ln 2}$$

$= 9.88\,\mathrm{cm}$ and $9.88/2 \approx 5\,\mathrm{bricks\ required}$

Q6. $\dot{D}_0 = (7.82 \times 10^{-8} \times 370)/0.5^2 = 1.16 \times 10^{-4}\,\mathrm{Gy/h}$

$= 1.16 \times 10^{-4}\,\mathrm{Gy/h} \times 30\,\mathrm{min} \times (1\,\mathrm{h}/60\,\mathrm{min}) = 5.79 \times 10^{-5}\,\mathrm{Gy}$

Q7. $37\,\mathrm{MBq} = 37 \times 10^6\,\mathrm{dps}, \therefore \left(37 \times 10^6\,\mathrm{days/s}\right) \times \left(1.12 \times 10^{-13}\,\mathrm{J/day}\right)$

$= 4.14 \times 10^{-6}\,\mathrm{J/s}$

$D_0^{\cdot} = (4.14 \times 10^{-6}\,\mathrm{J/s})/70\,\mathrm{kg} = 5.92 \times 10^{-8}\,\mathrm{Gy/s}$

Q8. $\bar{X} = 19980/5 = 3996; \quad s = \sqrt{287770/4} = 268.2;$

$\mathrm{CI} = (1.96 \times 268.2)/\sqrt{5} = 235$

Q9. Minimum counts $= (2.58/0.01)^2 = 66564$

Chapter 8

Q1. $\nu = 2.99 \times 10^8/670 \times 10^{-9} = 4.46 \times 10^{14}\,\mathrm{Hz}; \quad \tilde{\nu} = 1/4.46 \times 10^{14}$
$= 2.24 \times 10^{-15}\,\mathrm{Hz}^{-1}$

Q2. $\delta t = 1/(10^7 \times 2\pi) = 1.5 \times 10^{-8}\,\mathrm{s}$

Q3. $k = \left[\left(565\,\mathrm{cm}^{-1} \times 2.99 \times 10^{10}\,\mathrm{cm/s}\right) \times 2\pi\right]^2 \times 2.9465 \times 10^{-26}\,\mathrm{kg}$
$= 331\,\mathrm{N/m}$

Q4. $\nu\left({}^1\mathrm{H}^{37}\mathrm{Cl}\right) = 2990.6 \times \sqrt{1.6139 \times 10^{-27}/1.6163 \times 10^{-27}}$
$= 2998.4\,\mathrm{cm}^{-1}$

$\nu\left({}^2\mathrm{D}^{35}\mathrm{Cl}\right) = 2990.6 \times \sqrt{1.6145 \times 10^{-27}/3.1405 \times 10^{-27}}$

$$= 2144.3 \, \text{cm}^{-1}$$

$$\nu \left(^2\text{D}^{37}\text{Cl} \right) = 2990.6 \times \sqrt{1.6145 \times 10^{-27} / 3.1497 \times 10^{-27}}$$

$$= 2141.1 \, \text{cm}^{-1}$$

Q5. $\Delta E = 2.0023 \times 9.274 \times 10^{-24} \times 0.3 = 5.57 \times 10^{-24} \, J$

Q9. $u = \left(6.626 \times 10^{-34} \, \text{J/s} \right) \times \left(2.99 \times 10^{18} \, \text{s}^{-1} \right) / \left(1.67 \times 10^{-25} \, \text{kg} \right)$
$\times \left(2.99 \times 10^8 \, \text{m/s} \right)$

$= 39.8 \, \text{m/s}$

Chapter 10

Q9. $A = 96.2 \times \exp \left(-0.0105 \times 44 \right) = 61 \, \text{GBq}$ and

$^{99m}\text{Tc activity} = 0.957 \times 61 = 58.4 \, \text{GBq}$

Chapter 11

Q9. $n(\text{UO}_2) = 1000/270.03 = 3.70 \, \text{mol} = n(\text{UF}_6) = 3.70 \times 22.4 = 82.9 \, \text{dm}^3$

Appendix C

Units, Fundamental Constants, and Conversion Factors

SI Units

Length	l	metre, m
Mass	m	kilogram, kg
Time	t	second, s
Electric current	I	ampere, A
Temperature	T	kelvin, K
Force	F	newton, N, kg m s^{-2}
Pressure	P	pascal, Pa, kg m^{-1} s^{-2} (N m^{-2})
Energy	E, U	joule, J, kg m^2 s^{-2}
Charge	q	coulomb, C, A s
Magnetic field strength	B	tesla, T, kg s^{-2} A^{-1}

Fundamental Constants

Atomic mass unit	m_u	1.661×10^{-27} kg
Avogadro constant	N_A	6.022×10^{23} mol^{-1}
Bohr magneton	μ_B	9.274×10^{-24} J T^{-1}
Bohr radius	a_0	5.292×10^{-11} m
Boltzmann constant	k	1.381×10^{-23} J K^{-1}
Faraday constant	F	9.649×10^4 C mol^{-1}
Gas constant	R	8.314 J K^{-1} mol^{-1}
Nuclear magneton	μ_N	5.051×10^{-27} J T^{-1}

Planck constant	h	6.626×10^{34} J s
Rydberg constant	R_H	1.09737×10^5 cm^{-1}
Speed of light (vacuum)	c	2.998×10^8 m

Conversion Factors

Ampere	$1 \text{ A} = 1 \text{ C s}^{-1} = 6.2415 \times 10^{18}$ electrons s^{-1}
Atmosphere	$1 \text{ atm} = 760 \text{ mmHg} = 101325$ N m^{-1}
Barn	$1 \text{ b} = 10^{-28}$ m^2 $= 10^{-24}$ cm^2
Curie	$1 \text{ Ci} = 3.7 \times 10^{10}$ disintegrations s^{-1} $= 1$ Bq
Electronvolt	$1 \text{ eV} = 1.602 \times 10^{-19}$ J
	$1 \text{ eV g}^{-1} = 1.602 \times 10^{-16}$ Gy
Gray	$1 \text{ Gy} = 1 \text{ J g}^{-1} = 100 \text{ rad} = 6.242 \times 10^{15}$ eV g^{-1}
Year	$1 \text{ year} = 3.156 \times 10^7 \text{ s} = 8766$ h

1	2		3	4	5	6	7	8	9	10	11	12	13	14	15	16	17	18
hydrogen 1 H 1.0079																		helium 2 He 4.0026
lithium 3 Li 6.941	beryllium 4 Be 9.0122												boron 5 B 10.811	carbon 6 C 12.011	nitrogen 7 N 14.007	oxygen 8 O 15.999	fluorine 9 F 18.998	neon 10 Ne 20.180
sodium 11 Na 22.990	magnesium 12 Mg 24.305												aluminium 13 Al 26.982	silicon 14 Si 28.086	phosphorus 15 P 30.974	sulfur 16 S 32.065	chlorine 17 Cl 35.453	argon 18 Ar 39.948
potassium 19 K 39.098	calcium 20 Ca 40.078	57-70	scandium 21 Sc 44.956	titanium 22 Ti 47.867	vanadium 23 V 50.942	chromium 24 Cr 51.996	manganese 25 Mn 54.938	iron 26 Fe 55.845	cobalt 27 Co 58.933	nickel 28 Ni 58.693	copper 29 Cu 63.546	zinc 30 Zn 65.39	gallium 31 Ga 69.723	germanium 32 Ge 72.61	arsenic 33 As 74.922	selenium 34 Se 78.96	bromine 35 Br 79.904	krypton 36 Kr 83.80
rubidium 37 Rb 85.468	strontium 38 Sr 87.62		yttrium 39 Y 88.906	zirconium 40 Zr 91.224	niobium 41 Nb 92.906	molybdenum 42 Mo 95.94	technetium 43 Tc [98]	ruthenium 44 Ru 101.07	rhodium 45 Rh 102.91	palladium 46 Pd 106.42	silver 47 Ag 107.87	cadmium 48 Cd 112.41	indium 49 In 114.82	tin 50 Sn 118.71	antimony 51 Sb 121.76	tellurium 52 Te 127.60	iodine 53 I 126.90	xenon 54 Xe 131.29
caesium 55 Cs 132.91	barium 56 Ba 137.33	57-70	lutetium 71 Lu 174.97	hafnium 72 Hf 178.49	tantalum 73 Ta 180.95	tungsten 74 W 183.84	rhenium 75 Re 186.21	osmium 76 Os 190.23	iridium 77 Ir 192.22	platinum 78 Pt 195.08	gold 79 Au 196.97	mercury 80 Hg 200.59	thallium 81 Tl 204.38	lead 82 Pb 207.2	bismuth 83 Bi 208.98	polonium 84 Po [209]	astatine 85 At [210]	radon 86 Rn [222]
francium 87 Fr [223]	radium 88 Ra [226]	89-102	lawrencium 103 Lr [262]	rutherfordium 104 Rf [261]	dubnium 105 Db [262]	seaborgium 106 Sg [266]	bohrium 107 Bh [264]	hassium 108 Hs [269]	meitnerium 109 Mt [268]	ununnilium 110 Uun [271]	unununium 111 Uuu [272]	ununbium 112 Uub [277]		ununquadium 114 Uuq [289]				

57	58	59	60	61	62	63	64	65	66	67	68	69	70
lanthanum 57 La 138.91	cerium 58 Ce 140.12	praseodymium 59 Pr 140.91	neodymium 60 Nd 144.24	promethium 61 Pm [145]	samarium 62 Sm 150.36	europium 63 Eu 151.96	gadolinium 64 Gd 157.25	terbium 65 Tb 158.93	dysprosium 66 Dy 162.50	holmium 67 Ho 164.93	erbium 68 Er 167.26	thulium 69 Tm 168.93	ytterbium 70 Yb 173.04
actinium 89 Ac [227]	thorium 90 Th 232.04	protactinium 91 Pa 231.04	uranium 92 U 238.03	neptunium 93 Np [237]	plutonium 94 Pu [244]	americium 95 Am [243]	curium 96 Cm [247]	berkelium 97 Bk [247]	californium 98 Cf [251]	einsteinium 99 Es [252]	fermium 100 Fm [257]	mendelevium 101 Md [258]	nobelium 102 No [259]

Further Reading

The texts identified below will provide further information relating to the content of each chapter. Many of these books discuss topics in considerable detail, but will be useful for those requiring additional material. Undoubtedly, the reference work on nuclear chemistry is:

Vértes, A., Nagy, S., Klencsár, Z., Lovas, R. and Rösch, F. (Eds.) (2011) *Handbook of Nuclear Chemistry*, 2nd Ed. Springer-Verlag, New York, USA.

Chapter 1

Serway, R.A., Moses, C.J. and Moyer, C.A. (2004) *Modern Physics*, 3rd Ed. Brooks Cole, USA.

Young, H.D., Freedman, R.A. and Sears, F.W. (2003) *University Physics*, 11th Ed. Addison-Wesley, USA.

Chapter 2

Hanna, M.W. (1969) *Quantum Mechanics in Chemistry*, 2nd Ed. W.A. Benjamin Inc., USA.

McQuarrie, D.A. (1983) *Quantum Chemistry*, 1st Ed. University Science Books, USA.

Smith, B.E. (2013) *Basic Physical Chemistry: The Route to Understanding*, Revised Ed. Imperial College Press, UK.

Chapter 3

Krane, K.S. (1998) *Introductory Nuclear Physics*, 1st Ed. John Wiley & Sons, USA.

Chapter 4

Friedlander, G., Kennedy, J.W. and Miller, J.M. (1964) *Nuclear and Radiochemistry*, 2nd Ed. John Wiley & Sons, USA.

Chapter 5

House, J.E. (2007) *Principles of Chemical Kinetics*, 2nd Ed. Academic Press, USA.

Chapter 6

Choppin, G., Rydberg, J. and Liljenzin, J.O. (1995) *Radiochemistry and Nuclear Chemistry*, 2nd Ed. Butterworth-Heinemann Ltd., UK.

Chapter 7

Geary, W.J. (1986) *Radiochemical Methods*, 1st Ed. John Wiley & Sons, UK.
Peacocke, T.A.H. (1978) *Radiochemistry: Theory and Experiment*, 1st Ed. Wykeham Publications Ltd., UK.

Chapter 8

Banwell, C.N. and McCash, E.M. (1994) *Fundamentals of Molecular Spectroscopy*, 4th Ed. McGraw-Hill, USA.
Hollas, M.J. (2002) *Basic Atomic and Molecular Spectroscopy*, 1st Ed. Royal Society of Chemistry Publishing, UK.

Chapter 9

Malcolme-Lawes, D.J. (1979) *Introduction to Radiochemistry*, 1st Ed. Macmillan Press Ltd., UK.
Swallow, A.J. (1973) *Radiation Chemistry: An Introduction*, 1st Ed. Longman Group Ltd., UK.
Woods, R.J. and Pikaev, A.K. (1994) *Applied Radiation Chemistry: Radiation Processing*, 1st Ed. John Wiley & Sons Inc., USA.

Chapter 10

Attix, F.H. (1986) *Introduction to Radiological Physics and Radiation Dosimetry*, 1st Ed. Springer-Verlag, USA.
Saha, G.B. (2004) *Fundamentals of Nuclear Pharmacy*, 5th Ed. Springer-Verlag, USA.

Chapter 11

Atkins, P., Overton, T., Rourke, J., Weller, M., Armstrong, F. and Hagerman, M. (2010) *Shriver and Atkins' Inorganic Chemistry*, 5th Ed. W.H. Freeman & Co., USA.

Housecroft, C.E. and Sharpe, A.G. (2005) *Inorganic Chemistry*, 2nd Ed. Pearson Education Ltd., UK.

Chapter 12

Hore-Lacy, I. (2012) *Nuclear Energy in the 21st Century*, 3rd Ed. World Nuclear University Press, UK.

Lewis, E.E. (2008) *Fundamentals of Nuclear Reactor Physics*, 1st Ed. Academic Press, USA.

Mahaffey, J.A. (2012) *Nuclear Fission Reactors* (Nuclear Power), 1st Ed. Checkmark Books, USA.

Index